This publication has been reproduced to accompany the
Peebles-Richards Geneology

FOR MY MOTHER AND MY FATHER

Library of Congress Catalog
Card Number 76-50177
International Standard Book Number 0-87169-672-X
US ISSN 0065-9746

PREFACE

The Lonaconing journals, kept by the superintendents of the George's Creek Coal and Iron Company, cover the period 1837–1840. Their setting is the recent Appalachian frontier, and their subject is the building of an experimental iron furnace and the development of its adjacent company town in western Maryland. The principal figure in these activities was John Henry Alexander (1812–1867), topographical engineer for the state of Maryland, professor of mining and engineering at the University of Pennsylvania, one of the founders of the National Academy of Science, a member of the American Philosophical Society, and author and editor of works on many topics.

Alexander organized the coal and iron company fairly early in his career. Its furnace, auxiliary structures, and blast machinery were built to his design and erected under his supervision. The blast furnace was "experimental" in the sense that it was one of the first in the United States constructed specifically to use steam-powered hot blast with coke or raw bituminous coal as the smelting fuel. It became the prototype for other coke-fueled blast furnaces built in America before the Civil War.

The George's Creek Coal and Iron Company continued in business under that name until 1910, when its lands were sold to the George's Creek Coal Company. By some quirk of fate, the journals survived all moves, although masses of company papers were destroyed. One probable reason is that until 1910 descendants of the original incorporator were still active in the company and had a sentimental attachment to the early records. In turn, the president of the successor company (himself descended from an early settler in western Maryland) preserved the journal books in his private office, where they remained until his death. They are now held by his son, who has most generously permitted me to edit them.

I know of no other published or unpublished document which provides first-hand and generally nontechnical detail about the beginnings of an early nineteenth-century mining and manufacturing enterprise. The Lonaconing journalists touch on problems of supply, transportation, construction, labor, and discipline. They let us see stonecutters and masons at work, limeburners preparing kilns, brickmakers tempering and molding clay, and blacksmiths making tools and fixtures. They inform us of the opening of coal and iron mines, the laying of tramroads, the building of houses, the establishment of a store and post office. Throughout it all, displaying a not-unexpected condescension toward the working class, they try to impose the moral standards of Victorian gentlemen upon Welsh and German miners and artisans, and with the help of local magistrates deal with episodes of brawling, drunken roistering, and wife-beating.

I first saw these extraordinary journals nine years ago when their owner set down before me on a table in his home at Morristown, New Jersey, two canvas-bound volumes, each roughly 8 by 13 inches in size, and each containing, I guessed, about one-hundred pages of good watermarked paper faintly ruled in blue. The first glance at the content was discouraging; the initial pages bristled with mathematical calculations in minute, spiky handwriting. But as the journalist, John H. Alexander, began to develop his day-to-day record of events, I became absorbed in his account of the establishment of an ante-bellum iron works.

I had come to Morristown to do specific research for a book about Maryland coal miners. Within a few hours I decided that when I had finished *The Best-Dressed Miners* I would ask for permission to edit and publish the journals. I received that permission, together with a microfilm copy of the journals, in 1970.

Transcription of the microfilm resulted in 273 typed pages. Although I should have preferred to present the journals in their entirety, full publication seemed patently impractical in these days of rising costs of wages and materials. Some of the text is repetitious, and much of it consists of calculations, enumerations, surveyors' notes, and other detail of little interest to nonspecialists. Consequently, my first step in editing was to select material essential to a narrative of the progress of the furnace from ground-breaking to blowing in; material which would convey to the reader some sense of recurring crises—riots, strikes, fires, floods, landslides, and shortages of money, men, and supplies; material which would illustrate the problems of governing a work force of varied national origins and antipathies.

To the abridged text of the journal I have added editorial comments, explanations of construction and manufacturing processes, and identifications of persons and places. I have corrected misspellings and have adopted the American form of such words as "labor" and "molding." Each of the journalists had his own way of spelling certain proper names. I have arbitrarily chosen J. H. Alexander's versions because of his linguistic interests. I have written out abbreviations except in the case of a few proper names where it was not clear what was intended. I have not used initial capitals where they would not be used today. Some of the diarists used astronomical signs to indicate days of the week; I have omitted

these. However, I have indicated Sundays to give a point of reference. For the sake of the reader, I have occasionally added punctuation to clarify meaning. In no case have I altered the language of the journal.

ACKNOWLEDGMENTS

I wish to express my thanks to Lloyd R. Stallings for making the journal available to me and permitting me to edit it. Philip W. Bishop, curator of the division of manufacturing at the Smithsonian Institution, graciously allowed me the use of his library. Raymond Walker, deputy clerk of the Allegany County Circuit Court, was most helpful in the search for records at the courthouse. Matthew Skidmore, mine operator, transported me across rough country to the old Jacob Koontz house and cemetery. Alvin H. Ternent, descendant of an early storekeeper, was a source of Lonaconing lore and tradition. I am grateful also to the many other Lonaconing residents with whom I talked.

In addition, I am indebted to the staffs of the Library of Congress, National Archives, Maryland Hall of Records, Maryland State Library, Maryland Historical Society, Eleutherian Mills Historical Library, Maryland Academy of Science, the Archives Division of the Virginia State Library, Enoch Pratt Free Library, University of Maryland Library, the Vermont Historical Society, the Manuscript Division of the New York Public Library, Baker Library, Harvard University Graduate School of Business Administration, and Maryland Diocesan Archives, Protestant Episcopal Church.

K.A.H.

Silver Spring, Maryland
August 1976

FURNACE CHRONOLOGY

1837

Aug. 21. Working on foundation.
    29. East pier 5 feet high.
Sept. 4. South pier 5 feet high.
    20. Masons dressing arch stones.
    23. Masons covering alleys.
    29. Supporting beams placed over arches.
Oct. 2. Centers for arches placed.
    11. Masons at brickwork of arches.
    28. Brickwork of south arch finished.
    30. Hearth arch and east tuyere arch half finished.
        Preparation for first set of iron binders.
    31. Blacksmith preparing holdfasts.
Nov. 1. First holdfast laid.
    9. Epigraph of furnace set.
    10. North tuyere arch finished.
    11. Furnace height = 17'6".
    13. Blacksmith making screws for binders.
    14. Drying furnaces built in east and south piers.
    16. Second tier of iron binders laid.
        Drying furnaces built in remaining piers.
    21. Carpenter installs tram for describing interior curve of stack.
        Nuts and holdfasts put in for second set of binders.
    30. Drains laid in hearth.
Dec. 1. Bottom stone of hearth laid.
    2. Work on furnace suspended for winter.
    5. Boilers arrive.
    8. First load of steam engine parts arrives.
    18. Furnace covered to protect it from weather.
        Fires kindled in arches to dry furnace stack.

1838

Feb. 19. Foundations blasted for wing walls.
    25. Last load of steam engine arrives.
    26. Masons begin northeast wing wall.
Mar. 30. Masons begin southwest wing wall.
Apr. 2. Cover removed from furnace.
    9. Masons resume work on stack.
    11. Tram returned to interior of stack.
    21. Furnace stack 26 feet high
May 3. Stack 30 feet high.
        Fourth set of binders installed.
    16. Begin assembling steam engine.
    17. Stack at 36'6".
        Fifth set of binders installed.
June 4. Begin foundation for steam engine.
    9. Front of stack 50 feet high.
    21. Centers for bridge house arch put in.

    23. Begin bridge house arch.
    25. Finish foundation for steam engine.
July 7. Begin setting up frame of engine.
    11. Put steam cylinder on frame.
    16. Exterior of furnace finished except for flagging.
    23. Begin foundation for boiler furnace.
    30. Begin setting hearth stones.
Aug. 10. Dressing tymp stone.
    11. Tymp stone fitted.
    15. Masons fitting upper course of boshes.
    18. Finish foundation for steam engine chimney.
    22. Bricklayers making boiler furnace.
        Blast pipes placed in back alley.
    23. Begin foundation of bridge house.
    27. Lay foundation for regulator.
        Masons begin inwalls of blast furnace.
    28. Regulator placed in position.
Sept. 3. Boilers put in place.
    12. Begin foundation of south hot-air furnace.
    17. Bricklayers begin north hot-air furnace.
    21. Steam pipes put on engine.
    28. Boiler furnace finished except for flues to chimney.
Oct. 2. Blast furnace inwalls finished at top and being plastered with clay.
    3. All pipes in hot-air furnaces finished.
        Water pipes fitted to steam engine.
    5. Plastering of furnace inwalls finished.
    6. Boilers filled.
    8. Lay flagging at top of blast furnace.
    19. Blast furnace hearth complete.
    20. Boiler chimney finished.
    24. Fire lighted for drying hearth and interior of blast furnace.
        Fire lighted in boiler furnace.
    26. North hot-air furnace finished.
    30. Steam engine tried out.
    31. South hot-air furnace finished.
        Air pipes between engine and hot-air furnaces adjusted.
        Masons finish flagging top of blast furnace.
Nov. 6. Furnace hands place tuyere blast pipes.
    23. Water let in to boiler reservoir and boilers.
    24. Steam engine and blast machinery tried out.
Dec. 12. Adjustment of regulator completed.
    13. Satisfactory trial of engine and regulator.

1839

Feb. 21. Changes to hot-air furnaces completed.
     27. Hot-air furnaces tested.
Mar. 14. Contract for parapet and coping of blast furnace.
     23. Quarrying stone for bridge house floor.
Apr. 2. Begin boring pipes to supply water tuyeres.
     18. Laying water tuyere pipes.
     22. Blast furnace charged with coke and fired for drying.
May 8. Water tuyere pipes finished.
     9. Blast furnace blown in.
     16. Blast put on.
     17. First run-out of iron.

THE LONACONING JOURNALS: THE FOUNDING OF A COAL AND IRON COMMUNITY, 1837–1840

Edited by

KATHERINE A. HARVEY

CONTENTS

INTRODUCTION

On an April morning in 1837 two Baltimore gentlemen set out upon a short journey and a long venture. Boarding the Baltimore and Ohio train at Pratt Street, they headed west sixty-two miles on a circuitous route to Frederick, where they alighted six hours later.[1] The west-bound stage left Frederick at noon, passing through the fertile Middletown valley and over South Mountain at an average speed of perhaps five miles an hour to reach Hagerstown at five o'clock. In this tidy little village a comfortable inn offered overnight accommodation and food "prepared with abundance, and even elegance."[2]

Fifteen hours beyond Hagerstown lay Cumberland, the gateway to the West. Here stage drivers eased their coaches into a jostling procession of freight and emigrant wagons on the National Road, bound for Wheeling and the Ohio River steamboats. At the Narrows, a few miles beyond town, traffic squeezed through a pass between steep cliffs and began the long ascent to Frostburg, where our travelers left the turnpike and hired a livery-stable rig to take them to their ultimate destination, a narrow, sparsely settled valley eight miles to the southwest. Once arrived, they would be 150 miles from Baltimore—under the best conditions seventeen hours[3] away from family, friends, and their associates in a speculative mining and manufacturing enterprise. With only $30,000 promised or in hand, they were about to begin construction of an iron works estimated to cost when completed almost $168,000.[4]

The younger of the two men was John H. Alexander, son of an Annapolis merchant. Although his parents were by no means wealthy, they had managed to send him to St. John's College, where, at the age of fourteen, he had taken a degree in classics. After reading law for four years, he decided to become a civil engineer. At twenty-one he was appointed to make a study of the need for a topographical and geological survey of Maryland, and at twenty-two he became state topographical engineer, a position which he held until 1841. Although in early adolescence "extremely diffident, rather awkward, and retiring,"[5] he was described in his maturity as "tall, finely formed, erect, and easy in motion . . . exceedingly neat and precise in his dress," carrying himself always "with the air and bearing of a gentleman."[6] In 1837 he was twenty-five years old and newly married.[7] Already recognized as a competent engi-

---

[1] For contemporary accounts of travel over this route see Charles Varle, *A Complete View of Baltimore* (Baltimore, 1833), pp. 108–125; J. S. Buckingham, *The Eastern and Western States of America* (London [1843]) 2: pp. 137–160; Thomas Hamilton, *Men and Manners in America* (New York, 1968), pp. 157–163; and Frances Trollope, *Domestic Manners of the Americans*, Donald Smalley, ed. (New York, 1960), ch. *XVIII*. Travel time is from Buckingham. For later works based largely on the recollections of stage drivers and wagoners, see William H. Rideing, "The Old National Pike," *Harper's New Monthly Magazine*, Nov. 1879: pp. 801–816; and Thomas B. Searight, *The Old Pike* (Uniontown, Pa., 1894).

[2] Trollope, p. 201.

[3] An estimate based on a post office department advertisement (Hagerstown [Md.] *Mail*, Mar. 31, 1837) for proposals to carry daily express mail between Baltimore and Cumberland: 5 hours from Baltimore to Frederick (43 miles) and 9 hours from Frederick to Cumberland (91 miles). I have allowed 1½ hours for the uphill 10 miles between Cumberland and Frostburg and another 1½ for the 8 miles down the county road from Frostburg to the iron works at Lonaconing in the George's Creek valley. Ordinary mail service set up in 1839 allowed 7 hours by rail from Baltimore to Frederick and 14 hours from Frederick to Cumberland by four-horse post coach. Hagerstown *Mail*, May 31, 1839.

[4] George's Creek Coal and Iron Company, *George's Creek Coal and Iron Co.* (n.p., 1836), p. 30.

[5] William Pinkney, *Memoir of John H. Alexander, Ll.D.* (Baltimore, 1867), pp. 4–5.

[6] *Ibid.*, p. 28.

[7] On June 9, 1836, Alexander married Margaret Hammer, sole heir of Frederick Hammer, a Baltimore merchant, who

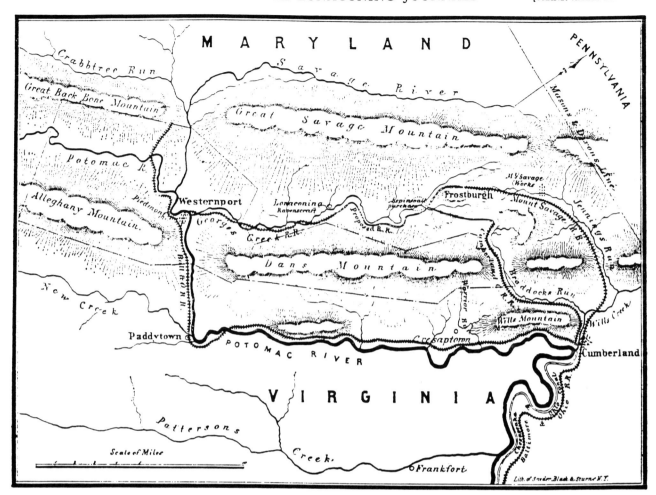

FIG. 1. Map of the George's Creek Valley, ca. 1855. All of the railroad lines were built after the period covered by the journals. Rankin, *Economic Value of the Semi-Bituminous Coal of the Cumberland Coal Basin.*

neer, he possessed also the imagination and the quality of persuasion characteristic of the antebellum man of enterprise.[8]

Alexander's companion, Philip T. Tyson, was thirteen years older than his friend and business associate. Trained in geology and chemistry, he was admirably suited to complement his partner in carrying out their western Maryland project.[9] The Tysons, an old Quaker family, had for several genera-

tions been engaged in manufacturing and trade in and around Baltimore. Philip's father, Isaac Tyson, Jr., was a pioneer industrial chemist, mining engineer, and metallurgist. At Plymouth, Vermont, he owned a water-powered charcoal blast furnace and a cupola furnace for casting. He had experimented with anthracite coal and hot blast for smelting copper ores, had obtained a patent for a hot blast stove, and was among the first in the United States to use hot blast for smelting iron.[10] He also owned and operated a copper mine in Frederick County, Maryland.[11]

In the valley to which the travelers came, the land and the waters were still unspoiled. Intricate patterns of conifers and hardwoods clothed the hills, and mulch lay deep on the forest floor, letting rainfall and melting snow seep slowly into the streams. In springtime, shadbush and dogwood blossomed in open

died in 1817 leaving an estate of roughly $75,000. As of Oct. 1, 1836, Alexander listed holdings in securities and real estate amounting to a little more than $28,000. Most of this property was a part of his wife's inheritance. Hammer Acct. Books, Vol. 1, folios 51–70, Md. Hist. Soc. This statement excludes bank deposits and money invested in western Maryland lands.

[8] For details of Alexander's career as scientist, linguist, and scholar, see Pinkney, *op. cit.*; John G. Proud, *Memoirs of Deceased Alumni of St. John's College, Annapolis* (Baltimore, 1868); J. E. Hilgard, "Biographical Memoir of John H. Alexander," *Biographical Memoirs of the National Academy of Sciences* 1 (Washington, 1877); and articles in *Dict. Amer. Biog., Nat. Cyc.*, and Appleton.

[9] Tyson obituary, Baltimore *Sun*, Baltimore *American*, and Baltimore *Bee*, Dec. 18, 1877; Appleton; *Nat. Cyc.*; and Tyson lineage and arms, Baltimore *Sun*, Jan. 22, 1905.

[10] Collamer M. Abbott, "Isaac Tyson Jr. Pioneer Mining Engineer and Metallurgist," *Md. Hist. Mag.* 60, 1 (March 1965): pp. 15–25; and Isaac Tyson Record Book, Md. Hist. Soc.

[11] Abbott, p. 21; Nancy C. Pearre, "Mining for Copper and Related Minerals in Maryland," *Md. Hist. Mag.* 59, 1 (March 1964): p. 22; and Hunt's *Merchant's Magazine* 5 (1841): p. 54.

woodlands, giving way in turn to pink and purple masses of laurel and rhododendron. As the year progressed, the glades blazed with a succession of showy plants—phlox, lilies, cardinal flowers, evening primrose, bee balm, and sunflowers. Trout abounded in the creeks and runs. Although deer were already becoming scarce, wild turkeys and pheasants were plentiful. The encroachment of settlers had driven panthers and bears into remote mountain fastnesses, where they were seen only occasionally. The most dangerous wild creatures were rattlesnakes and copperheads.[12]

By 1830 the presence of minerals in the Frostburg area was well known. Farmers had mined small amounts of coal since the turn of the century, and in 1828 the Maryland legislature had incorporated the first of the many companies which would exploit the region's coal veins on a larger scale.[13] Alexander and Tyson first visited the region in November 1833.[14] Three years later, in his capacity as state topographer, Alexander worked with the state geologist, J. T. Ducatel, and acquired first-hand knowledge of coal and iron deposits along George's Creek. Tyson, too, explored these deposits. All three men published studies on the geology and geography of western Maryland.[15] Moreover, through their membership in the Maryland Academy of Science and Literature, they met others who shared their interests, and they corresponded with such great figures as James Renwick and Benjamin Silliman, the latter particularly known for his work in chemistry and mineralogy at Yale.[16]

On the basis of their knowledge and experience, Alexander and Tyson pooled their resources and bought the tract known as Commonwealth, consisting of 3,817 acres along George's Creek at what is now Lonaconing, halfway between Frostburg and Westernport. The purchase price was $4,770;[17] the assessed value of this land in 1833 was $1,908.50.[18]

Preliminary exploration showed that Commonwealth contained promising veins of coal and of iron ore. With two or three additional partners, the new landowners could have financed a small charcoal furnace which, considering the growing demand for iron, would have given them a good return on their investment. However, what they had in mind was an establishment large in scope and advanced in technology—at least in America—and the young entrepreneurs decided to petition the Maryland legislature for a corporate charter which would permit them to raise further capital through the sale of shares.

An act of the General Assembly passed on March 29, 1836, granted John H. Alexander and Philip T. Tyson the right to form a corporation under the name of the George's Creek Mining Company with the authority "to open and work such mines of coal, iron and other minerals as may exist in the tract of land on George's Creek, called Commonwealth, now owned by them . . .," and to "erect and carry on mills and manufactories of iron." The act further allowed the company to acquire more land and to build a railroad from the company works to some point on George's Creek or on the Potomac River near the mouth of George's Creek. The capital stock was to consist of three thousand shares at $100 each in addition to the land owned by Tyson and Alexander, which would be considered part of the capital at a price to be determined by three disinterested persons—one appointed by the governor of Maryland, one by the proprietors, and the third by agreement of the other two.[19] A supplementary act passed June 1, 1836, changed the name of the company to the George's Creek Coal and Iron Company.[20]

During the six months following the granting of their charter, Alexander and Tyson acquired in their own names most of the acreage in Beatty's Plains, a tract southwest of Commonwealth. For this they made a partial payment of $3,031.10.[21] Purchase of Beatty's Plains increased the embryo company's real estate holdings to more than eleven thousand acres lying principally in the valley of George's Creek and along the eastern slope of Savage Mountain. The creek, running through company property, presumably would furnish the limited amount of power needed for the first stages of the contemplated operation.

Between March and September 1836 the owners made a more thorough examination of the topography and geology of Commonwealth. In October they published their findings, together with maps, in what they termed a "memoir,"[22] obviously hoping to attract investors. They reported that they had made a series of excavations at Dug Hill, a spur of Savage

[12] A description of Allegany county flora and fauna appears in J. T. Ducatel, "Outlines of the Physical Geography of Maryland, embracing its prominent geological features," *Trans. Md. Acad. Science and Literature* 1 (Baltimore, 1837): pp. 42–43.

[13] For a brief account of the development of the Maryland coal industry see Katherine A. Harvey, *The Best-Dressed Miners* (Ithaca, 1969), ch. 1.

[14] P. T. Tyson to J. H. Alexander, Nov. 22, 1864. Alexander Papers, Md. Hist. Soc.

[15] Alexander's and Ducatel's official reports to the governor of Maryland begin in 1834 and continue to 1840. See also J. T. Ducatel, "Outlines of the Physical Geography of Maryland"; and Philip T. Tyson, "A Description of the Frostburg Coal Formation of Allegany County, Md.," and "A Descriptive Catalogue of the Principal Minerals of the State of Maryland" in *Trans. Md. Acad. Science and Literature*, pp. 92–98 and 102–117.

[16] Membership of the Academy, *Trans.*, 1837.

[17] Allegany County land records, liber T, folio 298.

[18] *George's Creek Coal and Iron Co. v. Commissioners of Allegany County*, chancery records No. 8283, Maryland Hall of Records.

[19] Md. *Laws* 1835, ch. 328.

[20] *Ibid.*, ch. 382.

[21] Allegany land records, liber T, folio 327.

[22] Hereafter referred to as GCC&I Co. Report 1836.

Mountain near the southern end of Commonwealth, exposing the beds of coal and ore for a vertical height of five hundred feet. Their exploration led them to estimate that Commonwealth would yield almost 95 million tons of rich ore, practically free from sulphur and completely free from phosphorus, and nearly 158 million tons of coal suitable for smelting the ores. In sum, every acre contained, "after the most prodigal discount, the . . . ore and the fuel for making upwards of 9,000 tons of cast iron."

Besides coal and iron, the Commonwealth excavations revealed strata of good sandstone, limestone, and clay. It would therefore be unnecessary to go off the company's lands to find raw materials for constructing the furnace and for manufacturing iron. Heavy stands of white pine, oak, maple, walnut, hickory, locust, ash, and poplar would furnish lumber for industrial buildings, tramroads, and workmen's houses.

At the end of September, on the basis of personal inspection of the ground and consideration of the maps and reports contained in the memoir, the commissioners appointed to set a value on the company lands estimated them to be worth $800,000, assuming the subscription of $300,000 in money capital which would be applied to mining and manufacturing as contemplated by the company's founders.[23] Among the commissioners were Julius T. Ducatel, the state geologist of Maryland, and James Renwick, professor of natural philosophy and experimental chemistry at Columbia College, New York. Renwick, a nationally recognized authority on engineering, was intimately acquainted with the plans for development of the Alexander and Tyson holdings.

Even as late as 1836, no American ironmaster had adopted the whole set of technological advances in metallurgy which enabled British industry in the latter part of the eighteenth century to turn out both pig iron and wrought iron inexpensively and in great quantities. After 1815 some American establishments began to puddle their resmelted pigs to burn off impurities, and built mills for rolling iron bars and plates. But Americans were hesitant to make the basic change on which improvements in the secondary processes of iron making were founded. Despite the fact that furnaces in England and Wales had long since looked to coke for their smelting fuel, most American founders until well beyond 1840 continued to use charcoal, which produced a high-quality iron suitable equally for the blacksmith's forge, for the larger-scale manufacture of wrought iron, or for castings. Alexander was convinced that western Maryland possessed "great facilities . . . for a cheap manufacture with mineral coal."[24] He therefore

determined to make the experiment, and in 1839 his George's Creek Coal and Iron Company put into blast what may have been the first successful coke furnace in the United States, and the first in this country to use raw *bituminous* coal in the manufacture of iron.[25]

The use of coke was intended to be only the first of the company's innovations. The grandeur of the whole scheme may be seen in the detailed estimates which Alexander and Tyson secured from Professor Renwick. Plans for the manufacturing works at Lonaconing included four blast furnaces with steam-powered blowing apparatus and hot blast, together with molding houses, a foundry, a steam hammer, puddling furnaces, rollers, and a merchant mill. A furnace complex of this size would be capable of turning out seven thousand tons of wrought iron a year. In addition there would be a superintendent's house, four houses for managers and clerks, and accommodations for four hundred workmen and their families. All of this, plus an item of roughly $15,000 for "contingencies," would cost $167,930.[26]

Alexander, the guiding genius of the company, was aware that the growing demand for iron would soon outstrip the productive capacity of the American charcoal furnaces, few of which could run out more than a thousand tons a year. More iron would be needed for casting household goods, stoves, agricul-

---

[25] See Frederick Overman, *The Manufacture of Iron in all its Various Branches* (Philadelphia, 1850), p. 175; and James M. Swank, *History of the Manufacture of Iron in All Ages* (2d ed., Philadelphia, 1892), pp. 256, 369–370. The key word here is "successful," and its users did not tell us what constituted success. Overman said merely that the Lonaconing furnace in Maryland "was the first coke furnace, whose operation was successful, erected in this country." In discussing the Maryland iron industry, Swank wrote, "Overman claims that this [the GCC&I Co. furnace] was the first successful coke furnace in the United States." In a later chapter he stated, "The first notable success in the use of bituminous coal in the blast furnace in this country was achieved at three furnaces in Western Maryland"—one at Lonaconing, with which we are concerned, and two at Mount Savage, not built until 1840. In GCC&I Co's report to its stockholders in 1839, Alexander stressed the importance of "producing, as we have shewn it can be done, *iron with raw coal.*"

Certainly Alexander was not the first to try coke. Swank assigns that honor to an unnamed ironmaster at Bear Creek, Armstrong County, Pa. The furnace, built to use coke, was blown in in 1819, made a few tons of iron with this fuel, and because of insufficient blast was forced to change to charcoal. In 1835 William Firmstone made "good gray iron" for about a month at Mary Ann furnace, Huntingdon County, Pa., and in 1836 or 1837 F. H. Oliphant made a small quantity of iron with coke at his Fair Chance furnace near Uniontown, Pa. Oliphant did not continue to use coke, but soon resumed manufacture with charcoal, "probably from the higher value set upon charcoal iron." See Walter R. Johnson, *Notes on the Use of Anthracite in the Manufacture of Iron* (Boston, 1841), pp. 5–6; and Swank, pp. 367–368.

[26] GCC&I Co. Report 1836, pp. 27–30. It is interesting to note that the estimate employs the word "rollers" instead of the more commonly used "rolls."

---

[23] *Ibid.*, pp. 25–26.

[24] John H. Alexander, *Report on the Manufacture of Iron* (Annapolis, 1840), ch. III, p. ix.

tural tools, and machine parts. The railroads were already clamoring for supplies; the Baltimore and Ohio, for example, found that it could not secure from American mills the fifteen thousand tons of wrought iron straps necessary for covering its wooden rails. As a consequence, iron rails were among the principal imports from Great Britain, and they came in with a refund of duty once they were laid.[27] Alexander anticipated that with all four furnaces in blast, his company would produce a minimum of ten thousand tons of pig iron annually at a cost of $15 a ton and would sell it at the works for $25 a ton, making a profit of $100,000 a year on an investment of $160,000. If the pigs were converted to seven thousand tons of bar iron or railroad iron, the cost would rise to $30 a ton and the selling price at the forge to $60 a ton, allowing a profit of $210,000 for a year's operation. Market prices in 1836, according to the memoir, were higher than the assumed prices at the furnace, but there was a problem of transportation. Pig iron was selling for at least $35, and railroad iron was imported at $70 a ton,[28] but until completion of the Chesapeake and Ohio canal and the Baltimore and Ohio Railroad, the price at Lonaconing would have to be low enough to compensate for the high cost of hauling.

The profit from manufacturing was not the only inducement to investors. The memoir of 1836 pointed out that the sale of fifty thousand tons of coal annually would yield $52,000 over the cost of mining and transportation, and the carriage of coal for others would bring in $10,000, "a handsome interest" on the investment of $100,000 in the railroad which the company hoped to build.

By the terms of the original act of incorporation, once the lands had been valued the company might receive subscriptions to its capital stock, accepting a down payment of $10 a share. The balance was to be paid "as soon as conveniently may be" after the election of a president and directors.

Earlier in the 1830's it would have been fairly easy to raise $300,000. It was a time of heady optimism, with both English and American investors eager to put their funds into manufacturing, public works, and land. But by the autumn of 1836 a number of

factors presaged the coming panic of 1837. Investments became speculations, real estate in particular passing from hand to hand at rising prices in exchange for promises or pieces of worthless paper. Jackson's veto of a renewed charter for the Bank of the United States encouraged the proliferation of small-town banks which granted credit unwisely and issued huge amounts of unsecured banknotes. In turn, the Specie Circular of 1836, requiring payment for public lands in gold or silver, drew hard money away from the East. At about this same time, a credit crisis in England made merchants and bankers wary of accepting anything but specie from overseas debtors. The failure of several important British business houses flooded the London market with American securities, and it was indeed a persuasive Yankee who could dispose of his stocks and bonds in this atmosphere of financial uncertainty.

Alexander and Tyson had expected to sell much of their company's stock in New York and Baltimore. In 1837 the New York banks which held United States government deposits faced the crisis of distribution of surplus revenue which Congress had voted in 1836. Many of the moneyed men of Baltimore who would ordinarily have supported the business undertaking of a fellow citizen had already invested heavily in the Baltimore and Ohio Railroad or in the Chesapeake and Ohio canal, and the prospects of neither could at the moment be called bright. Under the circumstances it might have been prudent to ride out the economic storm before venturing into a sea of new enterprise. Therefore, it is somewhat surprising that only two months after publication of the memoir, the small group of stockholders who comprised the George's Creek Coal and Iron Company decided not to wait for further subscriptions but to organize and to begin operations.

One feels a sense of urgency in the correspondence between Thomas S. Alexander (oldest brother of John H.) and Dr. Patrick Macaulay, another of the early subscribers. Thomas, an Annapolis lawyer and an influential member of the Maryland House of Delegates, concluded that the company might go ahead with a minimum of $30,000, and he undertook to secure an act of the legislature to permit the company to proceed with this limited amount of money.[29] Behind his decision lay the need to be first on the scene in the development of Allegany County mineral resources. Nine other companies also possessed Maryland charters allowing them to mine coal and iron and to build works for the smelting and manufacture of iron in the western part of the state.[30]

---

[27] In 1832 Congress provided that duties were to be refunded on all imported rails laid down within three years of the date of importation. Frank W. Taussig, *The Tariff History of the United States* (8th ed., New York, 1931), pp. 56–57. As to the extent of imports, see Victor S. Clark, *History of Manufactures in the United States* (New York, 1949) 1: p. 511; Harry Scrivenor, *History of the Iron Trade* (London, 1854), p. 267; and Alan Birch, *Economic History of the British Iron and Steel Industry* (New York, 1968), pp. 220, 227.

[28] GCC&I Co. Report 1836, p 21. These estimated costs of production and profit are very close to those made in 1842, with benefit of the GCC&I Co. experience, by the Maryland and New York Iron and Coal Company for its own operations at Mount Savage, Maryland. See Maryland and New York Iron and Coal Company, *Statement of the London Committee of Management* ([London, 1842]), pp. 12–13.

[29] Thomas S. Alexander to Patrick Macaulay, Nov. 30, 1836. Alexander Papers.

[30] Allegany Iron Co., Md. *Laws* 1827, ch. 197; Md. Mining Co., *Laws* 1828, ch. 170; Savage Coal and Iron Co., *Laws* 1834, ch. 152; Union Co., *Laws* 1835, ch. 255; Town Hill Mining, Manufacturing and Timber Co., *Laws* 1835, ch. 306; Cumber-

Most of these, because of their location, would not interfere with the George's Creek Coal and Iron Company's plans, but one was a serious threat. This was the Union Company, one of the many enterprises of Duff Green, who began to develop coal and iron lands in Virginia and Maryland after his parting company with the Jackson administration.[31] Green's lands were near the mouths of the Savage River and George's Creek at Westernport on the Potomac River, and his charter entitled him not only to mine coal and to manufacture iron, but also to build railroads and to utilize part of George's Creek as the terminus of a canal from Cumberland.[32] In urging the raising of the $30,000, T. S. Alexander wrote:

We are at present in advance of General Green's Union Company and our railways will take precedence of his. But if we forfeit our present advantages it may be found that our company will become a mere appendage. If we should improve our present opportunities, we shall acquire an important influence, perhaps a control, over the whole valley of George's Creek.[33]

The supplementary act of the General Assembly passed on February 23, 1837, allowed the company to go into operation with subscriptions of a cash capital of at least $30,000, which must be paid in full within four months of organization of the company. If the amount were not paid within the time limit, the supplementary act would become null and void.[34]

Anticipating passage of the act, the company on February 10, 1837, gave public notice of a general meeting of stockholders to be held on March 4 to elect nine directors.[35] Since the charter required that the directors be stockholders, it is apparent that at least nine persons had already subscribed to or were otherwise entitled to shares of the capital stock. Alexander's fragmentary financial records[36] contain a list of twenty-two persons who were probably stockholders at the time of the meeting: J. H. Alexander, Thomas S. Alexander, William Alexander, Roswell L. Colt, Charles Augustus Davis, Thomas E. Davies, C. E. Detmold, Robert Graham, Patrick Macaulay, Louis McLane, Granville S. Oldfield,

Charles Oliver, Guy Richards, Morris Robinson, Dudley Selden, Philip T. Tyson, Isaac Tyson, Jr., James W. Webb, Samuel Wetmore, S. V. S. Wilder, Richard Wilson, and G. Leo Wolf. William Alexander, a Baltimore commission merchant, was a brother of John H. and Thomas. Guy Richards has not been identified. Dudley Selden, a New York lawyer, was briefly a member of the U. S. House of Representatives. Samson V. S. Wilder was New York agent for the French banking house of Henry Hottinguer and also agent of the Bank of the United States. The others are discussed elsewhere in the text.

How much ready cash had been raised is quite another matter. The company's lands, valued at $800,000, represented eight thousand shares. Of these, 7,019 eventually went to J. H. Alexander and Philip T. Tyson, and 981 to Charles Oliver of Baltimore who owned a substantial interest in Beatty's Plains.[37] The identity of certain "New York friends"[38] who were elected directors can be speculated upon but in most cases cannot be certainly established. As to the contributions which made up the $30,000, we know definitely that J. H. and Thomas Alexander each pledged $5,000.[39] Macaulay, Tyson, and the New Yorkers pledged unknown amounts.

In the matter of finance, and particularly in the solicitation of funds in New York, the George's Creek company depended to a great extent upon the business connections of Patrick Macaulay. Macaulay, a Baltimore physician, was president of the American Life Insurance and Trust Company, incorporated by the Maryland legislature in 1834.[40] Richard Wilson, secretary of the insurance company, was elected treasurer of the George's Creek Coal and Iron Company, of which he was a director, at the organization meeting in March 1837. By law, half of American Life and Trust's $2,000,000 capital was to be invested in public securities, bonds, ground rents, and mortgages, and the remainder in stocks and in real and personal securities. However, the company might not hold more than $20,000 in the stock of any one private corporation. It seems clear that American Life and Trust did invest in the George's Creek Company. In correspondence with Macaulay, Thomas Alexander referred to "the extent of assistance we might expect to derive from your office," and went on to say, "I now conclude that after we shall have

land Coal Mining Co., *Laws* 1835, ch. 334; Boston and N.Y. Coal Co., *Laws* 1835, ch. 342; Allegany Mining Co., *Laws* 1835, ch. 326; Clifton Coal Co., *Laws* 1835, ch. 365.

[31] In the bitter Jackson-Calhoun controversy, Green not surprisingly supported the latter. Calhoun's son was married to Green's daughter.
[32] Act of incorporation, Md. *Laws* 1835, ch. 255; and *The Coal and Iron Mines of the Union Potomac Co. and of the Union Co.* (Baltimore, 1840).
[33] T. S. Alexander to Macaulay, June 22, 1837, Alexander Papers.
[34] Md. *Laws* 1837, ch. 54.
[35] Advt., Baltimore *American*, Feb. 10, 1837.
[36] Alexander used the blank pages in vol. 1 of the Hammer Acct. Books (Md. Hist. Soc.) for his personal records and to a limited extent for noting early expenses connected with GCC&I Co. The presumed list of stockholders, written on a loose sheet of paper, was found in the acct. book at folio 93, which is headed "Office Accounts 1840."

[37] Oliver's subscription of land was authorized by Md. *Laws* 1835, ch. 382, sec. 6. See also Allegany land records, liber T, folios 380 and 483; liber W, folio 7, and liber AA, folio 346.
[38] T. S. Alexander to Macaulay, June 22, 1837. Alexander Papers.
[39] *Ibid.*, Nov. 30, 1836; and J. H. Alexander's Statement for the Arbitrators in the matter referred between himself and the George's Creek Coal and Iron Company (1850), Alexander Papers.
[40] Md. *Laws*, Dec. Sess. 1833, ch. 256. Supplementary acts 1835, ch. 269, and 1836, ch. 240.

raised $30,000 & applied that sum to the Land your office will make us advances."[41]

Perhaps more significant is Thomas Alexander's assurance to Macaulay, "Your acquaintance with monied men & with the monied market & your knowledge of the course of trade are of inestimable value to us. . . ."[42] The insurance company's New York branch was located in the office of the Bank of the United States,[43] and its trustees included men more than casually acquainted with Nicholas Biddle and his bank. One of these, Morris Robinson, was cashier of the New York branch of the bank from 1819 to 1836.[44] Charles A. Davis, one of Biddle's close friends, was a director of the State Bank of New York. Davis was also author of the Major Jack Downing papers concerning Andrew Jackson and the bank war.[45] Roswell L. Colt, an intimate personal friend of Biddle (and also of Philip Hone), was director of the Baltimore branch of the bank from 1816 to 1819. The Biddle-Colt friendship had further interesting ramifications. Biddle's wife was a cousin of Colt's wife, the former Peggy Oliver, daughter of Robert Oliver of Baltimore. Robert Oliver, a trusted Biddle adviser, was the father of Charles Oliver, whose Beatty's Plains lands were part of the George's Creek Coal and Iron Company capital.[46]

A tabulation in John H. Alexander's handwriting implies that Robinson, Davis, and Colt made or pledged "1st payments" of $750 each, and that Joseph L. Joseph and Samuel Wetmore, American Life trustees, paid or pledged $750 and $375, respectively.[47] J. L. and S. Joseph, who had a banking house at the corner of Wall and Hanover Streets, were agents of the Rothschilds.[48] Wetmore was a partner of Wetmore & Co., East India merchants.[49] Similar contributions are attributed to three other New Yorkers: Thomas E. Davies, real estate speculator;[50] James Watson Webb, editor of the New York *Courier and Enquirer;*[51] and John L. Crary, merchant. Crary's daughter Phoebe was married to C. E. Detmold, a German-born engineer who had much to do with the development of the iron industry in the 1840's.[52] By his own account Detmold was a stockholder of the George's Creek Coal and Iron Company in 1837 and served for a year on the board of directors.[53] The tabulation included the name of one non-New Yorker—Louis McLane,[54] president of the Baltimore and Ohio Railroad Company and a director of the Maryland Mining Company. McLane apparently made a first payment of $375.

Various American Life Insurance and Trust Company advertisements listing its trustees and officers indicate that Macaulay's acquaintance with "monied men" in New York City included Anthony Barclay, of the banking house of Barclay & Livingston; James Boorman, iron merchant; David Hadden, importer; Isaac Lawrence, president of the New York branch of the Bank of the United States; and lawyers Samuel Ruggles and John Duer. All of these might be considered potential investors in the George's Creek Company stock.

From one source or another, $30,000 had been subscribed by the middle of January 1837.[55] However, promises were not payments. Even Tyson and Alexander had not turned over the full amount of their pledges to the company, and as they jogged toward Lonaconing on the last stage of their journey, they had no certain knowledge that the money they needed would be paid before July 4, the deadline set by the legislature.[56]

[41] T. S. Alexander to Macaulay, Nov. 30, 1836, Alexander Papers.

[42] *Ibid.*, Dec. 10, 1836, Alexander Papers.

[43] Advt. Baltimore *American*, July 11, 1835.

[44] Shepard B. Clough, *A Century of American Life Insurance* (New York, 1946), p. 31; and Joseph A. Scoville, *The Old Merchants of New York City* (New York, 1862–1866) 1: p. 373, and 2: pp. 50 and 260.

[45] Nicholas Biddle. *Correspondence*, Reginald C. McGrane, ed. (Boston and New York, 1919), fn. 101, and letters 101, 257, 290, 292, 342. See also Arthur M. Schlesinger, Jr., *The Age of Jackson* (Boston, 1950), pp. 214, 277; Scoville, 1: pp. 84–85; and Moses Yale Beach, *Wealth and Biography of the Wealthy Citizens of New York City* (6th ed., New York, 1845), p. 8.

[46] Thomas P. Govan, *Nicholas Biddle, Nationalist and Public Banker, 1786–1844* (Chicago, 1956), pp. 36, 38, 62, and 76; and Biddle, fn. 1, p. 13, and fn. 2, p. 30. See also Philip Hone, *The Diary of Philip Hone, 1828–1851*, Alan Nevins, ed., (New York, 1927) 2: pp. 639, 742, 766, and 875.

[47] Undated memo filed under "Bills and Receipts 1839–1866," Alexander Papers. Davis's stockholding is confirmed in a letter to JHA dated 16 Apr. 1864, discussing the company's affairs at some length.

[48] The Josephs, who had been involved in the sale of Louisiana and Mississippi state bonds and in a New York real estate speculation, failed early in March 1837, and August Belmont took over the Rothschild interests. Govan, p. 307; and Fritz Redlich, *The Molding of American Banking* (New York and London, 1968), Part II, pp. 71, 333–335, 353.

[49] Scoville, 2: p. 293.

[50] *Ibid.* 1: pp. 132–134.

[51] Appleton.

[52] Scoville, 4: pp. 80–85. Crary is included in Edward Pessen's "The Wealthiest New Yorkers of the Jackson Era: A New List," *New York Hist. Soc. Quart.* 54 (1970): p. 158. Another of Crary's daughters was married to Dr. Leo Wolf, one of the presumed stockholders on Alexander's list of 22. Scoville, 4: p. 84.

[53] C. E. Detmold, *George's Creek Coal and Iron Co.* (n.p. [1849]), p. 1.

[54] McLane had been Jackson's Secretary of the Treasury from 1831 to 1833. When he refused to remove government deposits from the Bank of the United States, Jackson transferred him to the State Department. Bray Hammond, *Banks and Politics in America* (Princeton, 1957), pp. 345, 383–384, 413.

[55] T. S. Alexander to Macaulay, Jan. 13, 1837, Alexander Papers.

[56] *Ibid.*, June 22, 1837, Alexander Papers.

THE JOURNALS

1837

Throughout May and June, hoping for good news from Baltimore, Alexander and Tyson laid out their plans for beginning work. They had already decided to locate their iron works on a level area at the foot of Dug Hill where they would be close to the public road and would have access to building stone, limestone, iron ore, coal, and water. Tyson was to be superintendent with an annual salary of $2,500 plus house rent and fuel.[1] It was not foreseen that Alexander would spend a great deal of time at the works as a consulting engineer without compensation.[2]

At some point in their discussions the two men decided to keep a log, which they entitled "Memoranda of Calculations and Determinations for the Construction of the Lonaconing Iron-Works and Daily Transactions."[3] At intervals during June and July, Alexander set down "calculations and determinations" for a quarry railroad, for the sows of the furnace, the curve of the inside furnace stack, and the dimensions of bricks for the furnace arches. His notes indicate that he made use of formulae from Thomas Tredgold's *Practical Essay on the Strength of Cast Iron and Other Metals* (London, 1824). Entries of "daily transactions" began on August 21.

As Alexander and Tyson quickly discovered,

At such a distance . . . from the seaboard, and in a strange and thinly inhabited district, it was . . . very difficult to procure the force necessary for carrying [the works] on with the speed commensurate with our desires. Residences for the workmen had to be built; and neither the climate of that region nor its industrial products allowed of their being erected of the same temporary character and materials which on the public works in milder latitudes are allowable and usual. So far, too, as the materials for our purposes were concerned, though not distant, they were as yet inaccessible until roads and avenues could be cut to carry them.[4]

Depending in the main on contractors who would hire the men needed for constructing the various parts of the furnace complex, as well as the dwelling houses, store, church, sawmill, and other units which made up the village, the company was able to limit its own payroll. Similarly, it got along with a minimum of draft animals, arranging with local farmers for sleds, wagons, horses, oxen, and drivers for transporting stone and other building materials to the furnace and other parts of the works.[5]

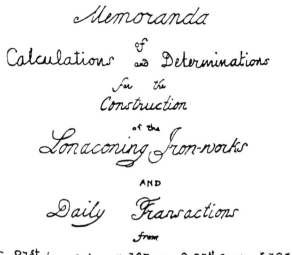

FIG. 2. Title page, volume 1, Lonaconing Journals.

So far as it is possible to determine from the journal, during 1837 the George's Creek Coal and Iron Company itself employed 10 carpenters, five blacksmiths, a wheelwright, thirty to fifty laborers, ten masons working on houses and preparing hearthstones for future use, a lime burner, and various quarrymen and miners, who may have been recruited from among the laborers. It appears also that some of the carpenters doubled as woodchoppers and hewers. A large proportion of this working force was made up of Germans from Baltimore, who found themselves quite at home in this region where many of the farmers were descendants of early German settlers.[6] A smaller number—miners, colliers, and furnace hands—were Welsh.

In professional or managerial positions, besides Alexander and Tyson, were the storekeeper, a chief clerk who also acted as postmaster, the "underground agent" in charge of the mines, and the founder. The company doctor, who arrived in mid-December, just barely fits into the category of first-year residents.

To accommodate its own workmen, as well as the various contractors and their hands, the company erected a number of boarding houses (referred to in the journal as "shantees") on the flat ground near the furnace. It also converted some non-residential structures to dwellings for employees who had brought their wives and children to Lonaconing. In addition, a few family men rented homes outside the boundaries

---

[1] George's Creek Coal and Iron Company, untitled report of directors, 1839 (hereafter referred to as GCC&I Co. Report 1839), p. 6, and J. H. Alexander statement of 1850, Alexander Papers, Md. Hist. Soc.

[2] Statement of 1850, Alexander Papers.

[3] Alexander called his settlement Lonaconing, the name supposedly given by the Indians to the George's Creek valley. GCC&I Co. Report 1839, p. 6.

[4] *Ibid.*

---

[5] As late as the end of 1838 it owned only 10 horses and had only a single wagon to send to town for goods. Journal, Dec. 21, 1838.

[6] Dieter Cunz, *The Maryland Germans* (Princeton, 1948), pp. 189–190. For example, the Fazenbakers mentioned in the journal were descendants of a Hessian prisoner of war who refused to be returned to Hesse, and settled in Allegany County about 1787. *Ibid.*, p. 191.

of the mining estate while awaiting completion of the log cabins which the company began to provide in September.

The earliest journal entries, showing the variety of activities already under way, reflect a preoccupation with internal logistics. Some rough cart roads had been cleared, and a railroad (tramroad) for bringing down sandstone blocks from a quarry on the hill above the furnace was operating, although the vehicles to be used on it had not been perfected. The most common form of tramroad of this period consisted of wooden rails supported on wooden cross ties. Wrought iron strips fastened to the upper faces of the rails gave them a durable surface.[7]

At the blast furnace, the excavation for the foundation had been carried down to solid rock; the hole had been refilled with large, hard stones, laid without mortar, with channels for drainage; and the masons were at work on the foundation itself.[8]

Except for its size, the furnace which Alexander had himself designed was not much different from the charcoal furnaces already operating in the eastern United States. The typical American blast furnace of the 1830's was seldom more than 30 feet high with a base 30 feet square. Alexander, following English and Welsh examples, drew plans for a coke furnace 50 feet high, 50 feet square at the base, and 25 feet square at the top.[9]

The outside of an antebellum furnace gives no hint of the massive interior construction. Supporting walls thick enough to withstand the heat and vibration of the blast surrounded the relatively small hollow interior in which the smelting process took place. At Lonaconing, for example, the interior had a diameter of $5\frac{1}{2}$ feet at the top of the stack (where the outer walls were 25 feet across), and $14\frac{1}{2}$ feet at the boshes, the widest part (where the outer walls were almost 50 feet across.)

To understand what was going on when the journal opens, one should visualize the outside of the Lonaconing furnace stack, when finished, as a truncated square pyramid with each of its lower faces broken by an arch sixteen feet wide. The archways, penetrating deep into the walls, tapered to a width of six feet at their inner ends. In the spaces not occupied

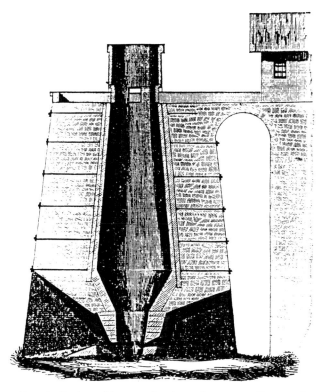

FIG. 3. Design of Lonaconing blast furnace as illustrated at Great Western Iron Works, Brady's Bend, Pa. Overman, *Manufacture of Iron.*

by the archways, masonry piers supported the upper part of the stack. The piers, on which the masons were at work on August 21, were made of large cut stones laid with mortar, and were solidly built to take the weight of the walls. Into the hearth and the supporting walls went blocks of stone weighing as much as 7,200 pounds apiece and measuring $6 \times 4 \times 2$ feet.[10] The only hoisting devices were windlasses and a wooden crane put up by the carpenters.

Journal entries by J. H. Alexander:

August 21 [Monday].—Working at the foundation. The road from the ore diggings progressing; one large tree cut up and removed. Quarry going on as usual: three blasts today and a portion to the left uncovered, out of which good stones come. Limekiln partially opened, but the stones not burnt, and therefore ordered it to remain till the night or the morning of Tuesday. Sleds hauling stone from the left of the foundation. Carpenter engaged in finishing the sled, which turns out too heavy for the rails, and the wheeled car was therefore put on and tried. The mare used for the railroad, and six men employed there for an hour. The wheeled car being as much too fast as the sled was too slow, directed at night a brake to be made for both wheels on the south side.[11]

[7] Frederick Overman, *A Treatise on Metallurgy* (6th ed., New York, 1882), pp. 86–87; and Charles Tomlinson, ed., *Cyclopedia of Useful Arts & Manufactures* (London and New York, 1854) 2: pp. 545–546. At Lonaconing the rails were made of oak 4 inches by 6 inches in cross section, and the metal strips were 2 inches wide and from $\frac{1}{2}$ inch to $\frac{5}{8}$ inch thick. Journal, June 22, 1837.

[8] Frederick Overman, *The Manufacture of Iron in all its Various Branches* (Philadelphia, 1850), pp. 153–164, describes in detail the construction of a blast furnace. See also Overman, *Metallurgy*, pp. 505–524; and Tomlinson, 2: pp. 77–78.

[9] John H. Alexander, *Report on the Manufacture of Iron* (Annapolis, 1840), p. 92; Overman, *Manufacture*, p. 175; and Alexander's statement of 1850, Alexander Papers.

[10] Journal, June 22, 1837.

[11] Heavy stones were moved along the rails on sleds. Lighter materials could be carried on the same rails on horse-drawn wagons. Where the incline was steep, wooden brakes were installed on the wheels on one side of the wagon. Tomlinson, 2: pp. 545–546.

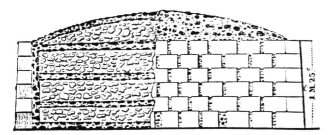

FIG. 4. First type of limekiln used at Lonaconing. *Dictionnaire de l'Industrie.*

August 23.—Great want of corner stones, and the sleds directed to procure them from above the foundation. Laid out a new road from the upper quarry for the sled. The wheeled car being found not to act [satisfactorily] in consequence of the too great height of the body, directed the carpenter how to lower it. Brickmakers here and commenced work on the eastern side of the creek.

August 24.—Rain last night and this morning. Directed the walls of the west piers to go up without batter [slope] (except the little which had been given to the west pier already) until six feet, when there will be a recess of 18" on which to rest the wing walls. [In the Welsh fashion, the Lonaconing furnace nestled into the bottom of a hill, with a wall at the back and "wing" walls at both sides to prevent the collapse of the excavation.] The sled road used this morning and four loads brought when one of the horses fell down and hurt himself. Directed a new road to be cleared and filled to give them a more gentle slope. The railroad car finished and tried [with] every promise of success. From some mismanagement, however, or misfortune, it broke loose at the upper end and ran off, breaking itself. Orders were given for the hands necessary (6 men) to be at the railroad on tomorrow morning early. The crane was also rigged.

August 25.—Commenced operations with the new sled at 5½ this morning, and matters went on successfully till 9, when the sled was permitted to break away and be broken up. Ordered a new one to be made, which in an hour was finished. Witnessed the satisfactory operation of the new sled, which delivers five loads of 21 cubic feet per load in one hour. Limekiln doing nothing but burning, and no mortar since 10 o'clock. George Blatter building a clamp for burning the limestone, which was lighted in the evening.

All of the mortar used for the furnace and other masonry was made with limestone quarried and burned on the premises. Alexander informs us in the journal entry for November 11, 1837: "Anterior to the opening of this register there had been three kilns started after the plan of M. Brard as described in *Dictionnaire de l'Industrie*, Vol. V. by M. Gaultier de Claubry.[12] As they did not turn out well, another plan was fallen on as recorded in the register of the 25th of August."

In Brard's plan an outer shell of sandstone blocks or bricks laid about a hand's breadth apart without the use of any mortar enclosed alternate layers of coal and limestone built up to the desired height. The illustration of his kiln shows five layers of coal and four of limestone. The bottom layer was of coal with fagots piled on top for kindling. The top layer was of fine coal covered with ashes, cinders, or earth to prevent loss of heat. The kiln was lighted at the bottom and allowed to burn until the stone had crumbled to powder—about thirty hours. After this the kiln had to cool for eighteen hours before the lime could be drawn out.

A clamp, unlike a kiln, did not have containing walls, the lime burner simply heaping his materials in a rough mound in the open. This does not mean that a clamp could be casually put together; the coal and limestone could not be thrown on in a higgledy-piggledy fashion if they were to burn properly. A sketch in the journal entry for August 25 shows that Blatter arranged the bottom coal in four rows approximately 6 feet long, 18 inches wide, 6 inches high, and a foot apart. His largest chunks of limestone applied carefully over the coal bridged the gaps between the rows and allowed the circulation of air from below. He then put on alternate layers of coal and limestone until he had constructed a crude pyramid roughly 4½ feet high. This particular clamp contained approximately 81 cubic feet of raw materials. Fires kindled in the spaces at the bottom of the pyramid ignited the coal, and the clamp burned until the limestone was reduced to powder. The journal notes the construction of 28 limekilns and clamps between August 1837 and April 1839.[13]

August 26.—Railroad too slippery this morning for more than one load. Hands therefore sent to the [ore] digging. The windlass finished and up by 10, delivering stones from the floor, which it does very well. Pauer procures about 4,000 feet of plank in Frostburg at 87½ cents per hundred. [Frederick Pauer was assistant engineer and chief clerk.]

August 27, Sunday.—P.T.T. [Tyson] arrives from Baltimore.

August 28.—Rupert dispatched to Frostburg for powder. [Rupert, properly Frederick Rupprecht, was a driver and general handyman who made special trips of this sort, transported visitors, and carried the mail.] The plank brought down today enough for the brick shed. Brickmaking going on well, and a small portion of the clay tempered. P.T.T. receives a letter from R.W. [Richard Wilson, secretary and treasurer of the company] announcing the readiness to pay up for Beatty's Plains [on which Alexander and Tyson still owed $11,128.43.]

August 29.—Foundation going on well. The east pier raised to 5 feet above the foundation. Carpenter directed to make the molds for the face brick of the arch wings. Calculation of the number of bricks for the arches postponed.

Everything possible was done to assist the brickmakers. Arriving late in the year, they lacked a

12 *Dictionnaire de l'Industrie, Manufactière, Commerciale et Agricole* (Paris, 1836–1841) 5: pp. 442–456. Cyprien-Prosper Brard (1786–1838) was a French mineralogist. For another contemporary description of limekilns see J. T. Ducatel, "Treatise on Lime-burning," in *Annual Report of the Geologist of Maryland, 1838* (n.p., n.d.), pp. 25–33. See also Tomlinson, 2: pp. 294–295; and Eli Bowen, *Off-Hand Sketches* (Philadelphia, 1854), p. 46.

13 The construction of clamps for burning other materials is discussed in Overman, *Manufacture*, pp. 43–44 and 119–121; Alexander, *Report on Manufacture*, pp. 144–145; and Overman, *Metallurgy*, pp. 289–290.

supply of clay broken up and mellowed by the frosts of the previous winter. In the remaining short season of 1837 they were assigned some of the company's laborers, who dug clay, threw it into shallow pits to be watered and soaked, and tempered it by treading and kneading it with their feet.

Working under a shed and on tables (molders' benches) made by the company's carpenters, the contractor and his hands began the shaping of the bricks. Each table was provided with a heap of sand and a trough for water. The molder laid on top of his table a pallet or barrow on which he placed a number of wooden molds without tops or bottoms. He dashed the clay into the molds, striking off any excess amount with a smooth piece of wood. He then removed the molds from the bricks, which were left to set on the barrow until they were firm enough to handle. As each barrow was filled, it was taken from the table and another was put in its place. Before using the molds again, the molder dipped them into sand or water.

When the bricks on the barrows were sufficiently firm, they were built up in long walls with space for ventilation on all sides, covered with straw to protect them from the weather, and left to dry until they were ready to be burned in a kiln. The burning process took about forty-eight hours.[14]

August 30.—Foundation stopped for want of mortar. Crane broke at 6 A.M., and two carpenters engaged through the day in repairing it. Broke with a stone of less than 2 tons.

August 31.—Foundation going on till about 10, when it commenced to rain and so continued during the day till past 5. Brickmaking behindhand today: 685 bricks molded before the rain. The scarcity of water, which had been a subject of complaint and investigation yesterday, is no longer felt. Fresh hands have to be sent tomorrow in order to keep up the tempering with the molding.

September 1.—Carpenter engaged all day in fixing the crane, which was barrelled up by battens [reinforced by strips of wood] and secured by two iron rings $\frac{1}{4}$ inch thick and fastened by screw-bolts. Foundation wholly stopped for want of stone, as neither [George] Stoup nor Geoghegan [farmers who did hauling] were here, and the crane could deliver nothing. One hand discharged for refusing to work in the brickyard. New plot of division of Beatty's Plains making,[15] and in the evening consideration of the molding house.

[14] Tomlinson, 1: pp. 186–189.
[15] Beatty's Plains had formerly belonged to Lewis Neth, Robert Oliver, and John Eager Howard, all deceased. Alexander and Tyson bought the Neth interest, $\frac{3}{4}$ of the tract, and Charles Oliver inherited his father's interest, $\frac{2}{3}$ of the remainder. The rest was divided among the Howard heirs. Tyson, Alexander, and Oliver brought an action in the chancery court to obtain a partition of the tract. In May 1837 chancellor Theodorick Bland appointed a commission of Allegany County residents to walk over and survey the tract, make a plat, and divide the land into 12 equal parts: 9 contiguous parts to go to Alexander and Tyson; 2 contiguous parts to Charles Oliver; 1 part to the Howard heirs. The commissioners were Thomas Shriver, Henry Koontz, Elijah Coombes, George W. Devecmon, and Robert Ross, Sr. Chancery Records, No. 9602, Md. Hall of Records.

September 2.—The two sleds hauling today, and the crane also delivering stone. The excavation for the south wall of the molding house finished as far as the sled road. The railroad put in operation with the new wheel car this morning and delivers in $1\frac{1}{2}$ hours 6 loads of 27 cubic feet each = 6 yards. This was at an expense of $18\frac{3}{4}$ cents per cubic yard, which is too dear. After deliberation, therefore, it was concluded to suspend operations there for the present.

September 5.—Corners of the first block of houses defined. In the evening three of the sows arrive.

The sows were heavy cast-iron beams, which would be placed across the archways to provide additional support for the walls above. In the final stage of making cast iron, the molten metal was run into trenches or depressions made in the sand floor of the molding house in front of the furnace. One of the legends of the iron industry is that pig iron is so named because these depressions were arranged in such a way that they resembled a sow suckling a litter of pigs. One long trench, the sow, led away from the tap hole at the bottom of the hearth. Short, narrow trenches, the teats, ran off at right angles to the sow, carrying the hot metal to roughly semicylindrical molds for the pigs. Because the connections between the pigs and the sows were thin, the pigs were easily broken off when the casting was cold. Both pigs and sows were crudely formed. It was possible, however, to produce uniform pigs by using wooden patterns to form the sand molds.[16]

The sows referred to in the journal at this point were specially ordered. Since Alexander had previously determined what the dimensions of the sows must be, it is probable that they were cast to pattern. Each weighed almost 3,500 pounds, and was 10 feet long, 12 inches wide, and $9\frac{1}{2}$ inches thick.[17]

September 6.—Raised the carpenter's wages to $1.50 per diem from 1st instant and agreed to give him when he removes his family [to Lonaconing] $1.625 and board or an equivalent. Agreed with Archibald Hook to come here and haul with a four-horse team and a two-horse team with two able-bodied drivers at $8 per diem. First death at Lonaconing—Hoehn's child. Selection of the site of the burying ground.

September 7.—Rain all day, and no out of doors work done. Consideration indoors of the centers for the hearth and tuyere arches, and of the water wheel.

The hearth arch, located at the front of the furnace, was also called the tymp arch or working arch. It was here that the molten metal ran out to the sow and pigs. Because several furnacemen were engaged in the casting of the pigs, and because they needed to move about freely, the hearth arch was substantially larger than the other arches. The tuyere arches gave access to the tuyeres, the cast-iron tubes at the end of the blast pipes, which directed a blast of hot air into the

[16] Tomlinson, 2: pp. 81–82; and J. E. Johnson, Jr., *Blast-Furnace Construction in America* (New York, 1917), p. 328.
[17] Calculations for the dimension of the sows of the furnace. journal June 26, 1837.

hearth of the furnace.[18] While an arch was being constructed, it was supported by a timber framework which also served as a model for the shape of the curve. This support was called a center.

The second "consideration" on this rainy Thursday reflects the original intention to have a 27-foot water wheel to provide power for the blast during most of the year and to use the steam engine as an auxiliary source of energy.[19] If the wheel was ever built, no trace of it remains today.

September 8.—Consideration of brick arches, and determination made of sizes as follows [with sketch of brick and dimensions of bricks and brick molds.][20]

September 9.—Foundation proceeding well, there being plenty of stone and lime. Level of the corners given at 1½ o'clock, and the south and west [piers] and hearth part of east pier found very close to the 5 feet whence the alley arches spring.[21]

September 10, Sunday.—Hook arrives with the last sow.

September 11.—The inner diameter of the hearth is determined at 34 inches and the number of segments (arch stones) at 8. Each segment then embraces an angle of 45° and the length of the arc is 13.35 inches. [The journal continues with calculations of the hearthstones, and sketches of the stones before and after dressing.]

Like the arch bricks, the hearthstones were wedge-shaped. Eight stones would be used to form each course of a circle, and so far as we can determine from the journal, the upper part of the hearth was enclosed by three courses of stone. Although the hearthstones would not be laid until much later, the preparation of these 24 stones involved so much work that it had to be started early. The dimensions worked out in the journal will give some idea of the problem involved in quarrying, transporting, and dressing the stones. As they came from the quarry, their parallel sides were almost 38 inches wide at the back and a little over 13 inches wide at the front. The slanting sides were 32½ inches long, and the stones were roughly 2 feet thick. In the final on-site dressing, the masons cut the straight front edge of each stone into an arc. Before any of this could be done, it was necessary to locate exactly the right kind of sandstone, free from iron and lime and able to resist great heat. The search began on September 11 with an order to uncover a new quarry for hearthstones.

[18] Overman, Metallurgy, p. 515; and Tomlinson, 2: p. 78.

[19] GCC&I Co. Report 1839, p. 7.

[20] Up to this date, the brickmakers had been molding ordinary bricks 4 × 8 × 2 inches, as was then the custom. The arch bricks, however, were beveled "to suit them to the proportions of the furnace." Journal, July 12, 1837. The sketch of Sept. 8 shows the burnt bricks to be 9 inches long. At the small end they were 4 inches wide and 2¼ inches high; at the large end, 4⅞ inches wide and 2¾ inches high. The molds were made slightly larger.

[21] Large furnaces had walkways or alleys through the piers so that furnacemen could go from one arch to another without having to go all the way around the outside of the furnace to check on the blast at each of the tuyeres and make adjustments if necessary. Overman, *Metallurgy*, p. 507.

September 12.—In the evening arrives Mr. Steele [who was to be in charge of mining operations.]

Alexander left Lonaconing for several days during which Frederick Pauer was in charge and made the journal entries.

September 13.—Started the digging for the wall of the engine house.

September 14.—The digging for hearthstone carried up to the cleared old field, but no good stone found. Tomorrow the ditch will be widened in some places to 4 feet, when I expect to find good stone. Stoup's and Giddigan's teams hauled stone. Giddigan himself did not come. I therefore pack his bag for him [discharge him] tonight.

September 15.—Engaged H. Coleman for Monday with 4 horses and wagon with stonebed. Bought 4,000 shingles from Green [whose lands adjoined Beatty's Plains], to be delivered in the course of next week at $6 per 1,000. Told [Frederick] Schmidt [who was in charge of laborers and haulers] to allow for breakfast only ½ hour and for dinner only 1 hour.

Alexander resumes the journal:

September 17, Sunday.—Fire bricks for lining the furnace to be 2⅝ inches thick. When burned the dimensions will be 10-5-4¼-2¼ inches. 4,600 will make the inner lining. 5,400 will make the outer course.

September 18.—Giddigan, who had been discharged, returns after proper concessions on his part, and engages to come again tomorrow. Stoup directed to go on with the cabins on the hill where the ground was marked out for him. He agrees to build the houses out and out except the chimneys, we finding plank for the floorings at the carpenter's shop, for $50 apiece.[22] The center line of the avenue [which was expected to be the town's main street] run and leveled. The entire avenue laid off into lots of 24 feet front. [Here the journal contains a sketch and surveyor's notes.] Mr. Steele starts a small coke heap.

Raw coal sometimes contained sulphur, which adversely affected the quality of the iron. Roasting the coal to produce coke removed the impurities. Like limestone, the coal was roasted in heaps in the open air. The workmen made an oblong pile of coal thirty-five or forty feet long and six or eight feet wide at the base, and four feet high. As in a limestone clamp, the draft was provided by the careful arrangement of large lumps of coal at the bottom. In some cases loosely built brick chimneys provided the draft. Like the limestone clamp, the heap was narrower at the top than at the bottom, but was mound-shaped rather than pyramidal. Coarse coal was piled loosely at the bottom and in the center of the heap; the rest of the pile could be made of coal in whatever size it happened to come. Altogether about thirty tons of coal went into the heap, which was then fired in three or four places around the base. The whole pile was covered with ashes and coke dust. In a few hours the fire reached the center, and the workmen used iron

[22] Eighty-five of these log houses were built. Testimony of Robert Graham, supt., in *George's Creek Coal and Iron Co. v. C. E. Detmold*, Chancery Records No. 8284, Md. Hall of Records.

bars to make air holes in the covering so that heat and smoke might escape. In four or five days the coke was ready for use. The Lonaconing coal yielded 76 per cent coke.[23]

September 19.—Coke turns out well. Contract made with Jacob Fazenbaker [a farmer whose land adjoined Beatty's Plains] to make 20,000 white pine shingles from Beatty's Plains and deliver them here at $5 per thousand. At night lengths of the holdfasts determined.

At predetermined intervals, the masons incorporated metal binders in the outer walls of the furnace to reinforce them and help to counteract the "active" expansion and contraction of the building materials with the heat of the blast. The binders, which were wrought iron rods or bars, went all the way through the outer walls from side to side of the stack. Large cast-iron washers (the holdfasts) slipped over the rods covered the binder channels and the adjacent stonework. The rods protruding through the washers were bent around into loops, and iron wedges thrust through the loops kept the binders in position. In some cases the ends of the binders were threaded so that large iron nuts could be screwed on. Both methods of fastening the binders were used at Lonaconing. The holdfasts, which were made by a Baltimore founder, weighed more than forty pounds apiece.[24]

September 20.—Mr. Steele engaged in reducing some of the iron, which was unsuccessfully prosecuted owing to the condition of the little furnace, which lacked draft.[25] A wooden trench or trough made for pounding the fire clay previous to sending it to [James] Totten, who agrees yesterday to grind it on his corn stones [at his grist mill]. The new carpenters arrive from Baltimore.

Tyson begins keeping the journal.

September 21.—Brick kiln filled and fired. Mr. Totten has ground 2 bushels of the fire clay, but finds it to make so much dust that he said he cannot go on with it. I suggested moistening it, and he will do so and try it again. J. H. Alexander left for Baltimore.
September 22.—Putting up a kiln for drying plank. James Sleman reports that the stones for the hearth will not split well.
September 23.—Mr. James Totten, Jr., has consented to accept my offer of three thousand dollars for his ninety-five acres of land [adjoining the Commonwealth tract to the south]. I have in consequence purchased it from him, and on Monday an agreement is to be signed. Masons nearly finished dressing stones for the alleys and recommenced covering them. Mr Schmidt clearing off the ore beds with 10 hands. Mr. Steele with some hands uncovering

the outcropping of the 14-foot coal and preparing for a drift into one of the upper ore beds.

The ore digging at Lonaconing during the summer of 1837 had been from deposits close to the surface, and did not require skilled miners. With a little supervision, laborers with picks, shovels, and wheelbarrows could remove the earth above a vein of iron ore and then take out the ore itself. Open workings were usually carried on in a series of terraces. This method of mining was inexpensive: one foot of stripping could be done for each inch of iron ore without involving the owners in high costs for wages.[26] After Steele arrived to take charge of iron and coal mining, Alexander and Tyson decided to carry some of the workings underground. Having previously determined where the ore was rich enough to justify the expense, they directed Steele to make a drift, or tunnel, into the hillside. Skilled workmen were required for this job.

September 26.—Mr. Totten has signed a memorandum of the sale of his land and has given up his smith shop. Mr. Pauer is having the shop repaired to set another hand at smith work. Bought Totten's blacksmith tools for $25.
September 27.—Mr. Steele is laying a rail track from the mouth of the proposed drift for ore. Clearing away for two sets of houses in order to set the masons more at work which have come to us today.
September 28.—A number of hands engaged in digging foundations for 4 stone houses [for managers and clerks]. Made an attempt to place the back sow, but the crane manifested such strong symptoms of giving way that we stopped.
September 29.—Hook's team dragged the sows up the road (leading up the hill above the furnace) to a level with the present altitude of the southern pier and then rolled them on the stack and fitted two of them on their places [over the arches] before dinner. Sugars, the new blacksmith, to make miners' tools under Mr. Steele's direction. Directed the wheelwright to make 4 tram cars for the ore and coal mines.

Tram cars used in iron mines were low in height so that they could be loaded easily. Generally they held from twelve to sixteen bushels, which, depending upon the quality of the ore, was the amount estimated to constitute one ton. Two axles, made of wrought-iron bars $1\frac{1}{2}$ to 2 inches thick, were fastened to the undersides of the wagons. The wheels were of cast iron, with or without flanges, depending on the kind of rail used.[27] The bodies for all tram cars at Lonaconing were made by the company carpenters and blacksmiths. The wheels, axles, boxes, and bearings were cast by Ross Winans, a Baltimore founder and designer of cars and locomotives for the Baltimore and Ohio Railroad.[28]

[23] Alexander, Report on Manufacture, p. 163; Overman, Manufacture, pp. 119-121; and M. Needham, The Manufacture of Iron (London, 1831), pp. 5-6.
[24] Daybook, Aug. 16, 1838, Welch and Alexander Record Books, Md. Hist. Soc.
[25] Steele was apparently testing the different types of ore, reducing the ore to metal in a small furnace about 2 or 3 feet high and 12 to 15 inches square at the base, with a flue leading to a chimney. Furnaces of this kind were usually built of firebrick. Overman, Metallurgy, pp. 180-181.

[26] Overman, Manufacture, pp. 59-60.
[27] Overman, Metallurgy, pp. 86-87.
[28] See Winans order book, Jan. 1839, and Winans journal entry Jan. 30, 1839, indicating that each set of four wheels, complete with axles, etc., cost about $19. Winans Papers, Md. Hist. Soc.

Tyson went to Baltimore on October 1, and Pauer began keeping the journal.

October 2.—Carpenters are engaged in putting up the woodwork for the arches. F. Pauer told the blacksmith, Michael, that he must work more steady if he intends to stay here.

October 4.—The brick kiln was opened today, but did not turn out well. More than ⅔ will not be fit for use. [William] Ledley [the brickmaker] promises to make good ones the next time. He excuses himself with the wood being too green and he having been hurried too much.

October 5.—Stoup does not push on with the cabins on Dug Hill as he promised to do. F. Pauer told him that they must be done by Saturday next week, which Stoup promises to do. Michael, the blacksmith, was discharged.

October 7.—The weather turned very cold, and the night frosts are too severe. The clay diggers were taken away from the brickyard and sent to the railroad digging.

October 9.—As the nights are so cold, the hands complained, and the masons agreed to do the lining of their boarding house [with bricks] in two nights. Brick was hauled to [Edward] Zacharias' shantee, any how not fit for the arch work on the furnace. The putting up of a scaffold for the stone houses was begun this afternoon. F. Pauer made a valuation that up to this date the perch [approximately 25 cubic feet of stone work] costs about $1.12½. The new blacksmith is doing well.

October 10.—The papers about the new post office at Lonaconing arrived from Washington.[29]

October 11.—[Charles] Pagenhart [a carpenter] was told to leave on no account the house which he rented from [Samuel] Buskirk. He will be sustained by force if necessary, according to the advice of the counsel of the company.[30] Three of the masons are laying brick. F. Pauer made a calculation that there are wanted about 16,000 bricks more.

October 13.—Three of the masons did begin to line Lohider's shantee with brick.

October 16.—Ledley says the brick [kiln] is not cooled enough to open it. We must leave it till tomorrow. It delays the work on the furnace a good deal. [Carpenters] engaged up to this time mainly in making window frames and the doors for the stone houses, also in ripping plank for the floors.

October 17.—The two cabins at Dug Hill built by T. Clarke are done. On Saturday two families will move up. The other two will be finished by Tuesday next. T. Clarke will build the four chimneys for $18 the piece. No bargain was made by F. Pauer, as Mr. Tyson or Mr. Alexander will be up here certainly by Monday next.

October 18.—Lohider is hauling timber for the molding house as fast as Hoehn and Pagenhart can hew it. The larger timber has been hewed by [John] Arner, and he will frame [the molding house] now. The brick kiln was opened yesterday and turns out to be well. About ¾ of the brick will be fit for the arches. A new brick kiln of about 14,000 will be fired tonight.

October 19.—Four new hands were engaged today. One of them is a blacksmith. Carts are much wanted, but not to be got, neither at Frostburg nor Cumberland. The

rails for a railroad track in the coal mine are partly laid today. Mr. [David] Hopkins, the new founder arrived. Mr. Tyson thinks he is a first-rate founder, well recommended to him [by William Kemble, of the West Point Foundry].

October 20.—Mr. Hopkins took charge of all the furnace affairs, as directed by Mr. Tyson. He seems to be much pleased with the prospects here.

October 21.—Two of the families moved up to Dug Hill today. Made an agreement with them to build their chimneys, $6 allowed per chimney.

October 22, Sunday.—The team of Oliphant's Iron Works arrived [on October 16] with the iron, freight 50 cents per hundred.[31] A wagoner from Baltimore arrived here with store goods and stoves, also some steel and grinding stone for the company.

October 23.—The hands working at night on the coal mine got promised $1.18¼ cents per night including the oil they use.

October 24.—Hopkins proposes to have the hearth five feet instead of three. Postponed till the arrival of Mr. Tyson or Mr. Alexander.

October 25.—The third brick kiln was opened yesterday. The brick seems to be as good as the last kiln. According to a calculation made by F. Pauer, this kiln ought to turn out eleven thousand brick to finish the arches.

October 26.—Both Sugars (blacksmiths) left the place without giving any notice, nor having any reason for it.

Alexander resumes the journal:

October 27.—P. T. Tyson and J. H. A. return today from Baltimore at 12 o'clock and find the work, owing to the operations at the bricks, not to turn out so far advanced as they had expected. The new houses on the avenue had attained under the charge of three masons a height of 10 feet from the ground. The ore digging which had been cleaned out was in the condition Mr. Tyson had left it. No drift had been carried in nor any open work further commenced. A small part of the frame for the molding house seems to have been finished. The back arch of the furnace was closed in, and four bricklayers working at the southwest arch. The centers of the others and the seats were all in. Mr. Hopkins appears well satisfied with his prospects, but there had been some difficulty from mis-efficiencyness [sic] in Mr. Steele. The railroad was not quite half done, and the coal mine just beginning to be entered into, a roof being over it about 75 feet in. Fazenbaker had made no shingles.

October 30.—Hearth arch about ½ done in all, and the east tuyere [arch] about the same. Preparation for laying the iron binds. Mr. Steele dissatisfied with his situation and told to be as easy as he could be.

October 31.—Iron ore with seven hands advancing. Disturbance among the masons and Zacharias [storekeeper and operator of a boarding house], but appeased. Blacksmith engaged in welding the iron holdfasts for the furnace.

November 1.—One of the holdfasts laid.

November 2.—Furnace going on slowly. Mr. Harris [mason contractor] doubts the practicability of getting done this winter. Carpenters engaged in refitting the school house for Arner's dwelling. Plan for Pauer to take Totten's house given up, and determination made to put in Steele. The building of the new houses abandoned,

[29] The Lonaconing post office was established officially on October 5, with Frederick Pauer as the first postmaster. Postal Records, National Archives.

[30] The Buskirks had been resident in the valley since 1788. Their lands adjoined Commonwealth, and GCC&I Co. seems to have maintained a running feud with the family concerning property boundaries. Possibly the house in question was on land claimed by the company. In various county records the family name is spelled Van Buskirk or Vanbuskirk.

[31] Most of the iron for the blacksmiths came from F. H. Oliphant, who had a charcoal blast furnace, forge, rolling mill, and nail factory at Fairchance, near Uniontown, Pa., about 60 miles by road from Lonaconing. His iron and nails had a wide reputation for excellence. History of Fayette County, Pa., Franklin Ellis, ed. (Philadelphia, 1882), pp. 231–232, 236–239, 583–587.

and an announcement made to the hands that they shall be allowed (masons) $20 apiece to finish [at the furnace] by the 5th December.

November 3.—[Twelve hands] making a new road to the furnace so as to deliver stone from the sleds. At horn blow nearly done. One man injured in the morning by the fall of a tree. Fire clay beating going on through the day, and in all about 230 bushels ground.

November 4.—[August] Weisskettle [stonecutter] shown the boundary stones for the Commonwealth tract and directed to letter them G.C.Co.

November 6.—J.H.A. and P.T.T. take a walk to Anderson's [a neighbor] to engage bricks from him, which they do upon condition that they replace them next summer when they are wanted. Ledley wants, however, $4.50 per thousand for making them through the winter and spring, which was decided to be too much. The brickmakers discharged today. Two new stonecutters engaged and commence work at the furnace. Also the hands from the houses transferred there except one who hurt himself. At night letter to Trimble in regard to miners.

The company needed skilled miners for underground work, and wished to secure some from Wales. The letter was probably addressed to Isaac Trimble, chief engineer for construction of the Baltimore and Susquehanna Railroad, who regularly bought rails from the Dowlais Iron Company at Merthyr Tydfil.[32] Trimble, like Alexander and Tyson, was a member of the Maryland Academy of Science and Literature.

November 7.—Early in the morning difficulty among some of the German masons, which was not appeased till the forenoon late, by which $\frac{2}{3}$ day was lost. At night August Weisskettle cutting the epigraph of the furnace, to be laid over the key [stone] of the tymp arch.[33] A new team of horses engaged. Totten settled with today, and a bond given for $3,000 with interest from 4th November. Deed for the Totten purchase sent to [William] Matthews [the company's attorney at Cumberland] to be recorded.[34]

November 8.—Letter from Mr. [George W.] Rodgers [Chesapeake and Ohio canal agent] in relation to the boilers, which have been at Williamsport since the 24th October. Letter written to Kemble about the engine.

In addition to bricks, stone, and mortar, building an advanced type of iron furnace involved a large amount of machinery for providing the hot blast necessary when coke was used as the smelting fuel. For this purpose the George's Creek Coal and Iron Company ordered a steam engine, boilers, and blowing cylinder from the West Point Foundry Association machine shops in New York City.[35]

It was originally intended that all of this machinery, an estimated twenty tons, should be sent by ship from New York to Georgetown, the head of navigation on the Potomac River and the eastern terminal of the Chesapeake and Ohio canal. At Georgetown the machinery would be unloaded and put on canal boats for carriage to Williamsport, a shipping and transfer point west of Harpers Ferry and about nine miles directly south of Hagerstown, Maryland.[36] In 1837 the canal boat charge for carrying castings and other manufactured iron between Georgetown and Williamsport was $3.50 a ton.[37] As it happened, only the five boilers were shipped in time to arrive in Maryland before ice closed the canal. Each of the boilers was twenty-four feet long and thirty-six inches in diameter.[38]

November 9.—Letter from Matthews enclosing the Commonwealth deed, which requires the addition of the clerk's certificate to the magistrate's.[39] Weisskettle sets his epigraph. [William Shaw] is engaged to go to Williamsport for the boilers with four teams.

William Shaw, who is mentioned frequently in the journal, was the son of one of the earliest settlers in the area. He was probably the most important resident landholder in the George's Creek valley, owning several thousand acres running on both sides of the creek from the southern edge of the company's property down stream almost to Westernport. He had a "mansion house," a sawmill, and a gristmill at what is now Moscow, four miles below Lonaconing. The governor of Maryland appointed him a justice of the peace in 1828 and a magistrate in 1836.[40]

November 10.—At night letter of W. Alexander, advising us of the transmission of a box of screw taps for the binder ends.

William Alexander, a brother of John H., was a commission merchant in Baltimore. In addition to operating the store at Lonaconing and providing the stock for it, he acted in general as a factor for the

[32] Madeleine Elsas, ed., Iron in the Making: Dowlais Iron Co. Letters, 1782–1860 (London, 1960), pp. 144–145.

[33] The epigraph is a small stone bearing the inscription "G.C.C. & I. Co./No. 1/J. N. Harris/1837." It is still legible, although vandals have destroyed part of the lettering.

[34] The deed for the Totten property was recorded Nov. 9, 1837, Allegany County, Md., land records, liber T, folio 387.

[35] GCC&I Co. Report, 1839, p. 7. The reference in the journal is to William Kemble, younger brother of Gouverneur Kemble, founder and president of the Association, which manufactured ordnance at Cold Spring, New York. When steam navigation became established on the Hudson, the Association erected machine shops in New York City to manufacture steam

engines and boilers. William Kemble was in charge of this branch of the business. Collections N.Y. Hist. Soc. 17, Kemble Papers (New York, 1885): p. xxi, prefatory note. See also Dict. Amer. Biog., and William J. Blake, History of Putnam County, N.Y. (New York, 1849), pp. 239–246.

[36] J. H. Alexander to J. P. Ingle, Aug. 7, 1837. Chesapeake and Ohio Canal Records, National Archives.

[37] Advertisements of William Holliday and Joseph Hollman, Hagerstown [Md.] Mail, Mar. 10, 1837.

[38] Walter R. Johnson, Notes on the Use of Anthracite in the Manufacture of Iron (Boston, 1841), note p. 8. Johnson visited Lonaconing in June 1839.

[39] On Oct. 5, 1837, Alexander and Tyson, in exchange for stock, transferred all their interest in Commonwealth to GCC&I Co. Deed recorded Oct. 26, 1837, Allegany land records, liber T, folio 380.

[40] James W. Thomas and T. J. C. Williams, History of Allegany County (n.p., 1923) 1: pp. 640–641; 2: p. 1094; and J. Thomas Scharf, History of Western Maryland (Philadelphia, 1882) 2: pp. 1354 and 1466. See also Allegany County land records. The list of patents and deeds is too long to be cited here.

company, seeing to it that materials for the furnace were ordered and shipped. The screw taps mentioned here were made by Charles C. Reinhardt, a Baltimore instrument maker, who charged $32. They were sent to Frostburg on the Stockton & Stokes stage at a cost of $4.[41]

November 11.—The furnace may be considered at 17′ 6″ high. A new fixture made for the mortar by which it can be carted up from the lime yard and shot down on the level of the upper sled road to the stack.
November 12.—Rain which began at 3½ yesterday and continued through the night with great violence was found this morning to have done considerable damage. Limekiln was extinguished, and divers washings about the furnace.
November 13. Blacksmith making the screws for the binders. The carpenter receives his directions about the top house, which is to be 33 feet long in the clear and 24 from out to out wide. [Further specifications are listed here for floor chords and roof.]

The top house was also sometimes called the bridge house. The top of the furnace (where it was charged) and the top of the hill against which the furnace was built were connected by a bridge made sometimes of wood, sometimes of stone or brick. A frame building on this bridge served as a storehouse for the fuel, ore, and flux needed to feed the furnace for one or two days. The roof extended over the top of the furnace, protecting it also from rain and snow.[42]

November 14.—In the morning determination for construction of the inner curve of the furnace [with sketch and calculations].[43] Masons do not work in the morning until after dinner for the snow and rain. Screw tap of Reinhardt breaks—the largest one—and blacksmith employed in making another. In the afternoon some disturbance among the miners under Mr. Steele, and four of them discharged. Four of the masons employed in building the [drying] furnaces in the east and south piers, which, however, they do not finish. At night Aurora Borealis.
November 15.—Difficulties with the wagoners in inducing them to take coal down [to Baltimore]. One sled engaged in hauling coal for them.

From time to time coal was shipped to the households of company officers and directors. However, many drivers preferred to go back to Baltimore with empty wagons rather than carry dirty loads for which they would receive only sixty cents per hundred pounds. William Alexander, who received the coal and paid the freight charges, wrote to J.H.A. on one occasion, "The fellow that brought the Doctor's [Macaulay's] I think watered it and by that means increased the weight, for if I am right it weighed 90

odd pounds per bushel. The wagoners are most able to cheat when they please in some way."[44] A bushel of Lonaconing coal was assumed to weigh eighty pounds at the pit mouth.

November 16.—William Shaw reports ill success in procuring wagons. Stack progressing, and the second tier of binders laid after dinner. Roads in such bad order that four sleds do not deliver fast enough for the crane. Letter to W. Alexander, enclosing his account for the proposed transaction in purchasing out his store [at Lonaconing].[45]
November 18.—Mr. Schmidt goes to Cumberland about a team, which he purchases. John Thomas and John Phillips, hands of Hopkins, arrive. [Like Hopkins, both were Welsh.]
November 20.—Phillips and Thomas, the two keepers who arrived on Saturday night, exhibit themselves on the premises.

Next to the founder, the keepers were the most important of the workmen who dealt with furnace operations. A furnace of moderate size usually employed two keepers, who were responsible for determining how much ore and coal should go in at the time of charging, and how the blast should be regulated. If the founder himself did not tap the furnace when the smelting was finished, the keepers performed this operation.

November 21.—Tram for describing the interior curve of the stack put up. Holdfasts and nuts put in this evening. Hands engaged in dressing hearthstones by torch light.
November 22.—Stack progressing well until 1 o'clock, when a storm came up, and the major part of the masons declined working during the evening [afternoon]. The centers [for the arches] all slacked after the rain had ceased. Letter to G. W. Rodgers in relation to the boilers, William Shaw having announced that the wagons were all obtained and ready to start. Violent snow storm at night with lightning.
November 23.—The bottom stone of the hearth hauled today from the ore digging and deposited in the molding house flat.
November 24.—Extremely cold today and snow spits through the morning. The large hearthstone hauled up by means of a windlass in the hearth and the aid of 10 men below with handspikes. A shed erected over the front of the tymp arch in order that the hands may dress the stone at night, which they do. The tymp [arch] centers also renewed. The drawing of the patterns for the second tier of hearthstones finished, as also of the tymp stone. Stoup and Co. engaged a greater part of the day do not get one stone down, and at night he is discharged and direction given to Mr. Pauer to employ him no more.
November 25.—The promise of fine weather given last night is belied by clouds and spits of snow at intervals through the day. Mr. Schmidt is sent to superintend

[41] Wm. Alexander Acct. Book, Nov. 8 and Dec. 15, 1837, Welch and Alexander Record Books, Md. Hist. Soc.
[42] Overman, *Metallurgy*, pp. 510, 524.
[43] The sketch shows the inner curve as beginning 12 feet above the hearth and continuing up for 32 feet. It diminishes from the "semidiameter" (radius) of the boshes—7 feet—to the "semidiameter" of the trunnel head—"2 F. 75." The space for the seat of the inwalls is "2 F. 33."

[44] W.A. to J.H.A., Nov. 25, 1837, Alexander Papers. For freight charges, see daybook entry, Nov. 30, 1837, Welch and Alexander Record Books.
[45] William had two weeks earlier expressed his willingness to let the company take over the store. W.A. to Tyson, Nov. 6, 1837, Letter Book 1, Welch and Alexander Record Books. The company also apparently took over Edward Zacharias, the storekeeper.

the bringing down of the hearthstones, which he does very satisfactorily, 5 being hauled to the edge of the hill above the furnace after nine o'clock.

November 28.—Work continued on the back wall, the molding house wall, and hearthstones. Hoepfel sent to split out the stones of the third course.

November 29.—Mr. Oldfield arrives this morning.[46] Day occupied in shewing him the premises.

December 1.—Workmen at the coal put at wages of 25 cents per cubic yard of coal got out, exclusive of fine coal, with an addition of 25 cents for every yard forward. Work at the stack on taking down the tram and bracing the crane in order to bed the stone of the hearth bottom.

The hearth, the lower part of the furnace where the heat was greatest and where the smelting of the ore was completed, was built of large slabs of sandstone cemented with fire clay. The bottom was formed of one large stone (referred to as the bottom stone) supported by a mass of masonry in which channels were left open to allow the escape of moisture. A typical bottom stone would be from twelve to fifteen inches thick, at least four feet wide, and six feet long.[47]

December 2.—[John] Harris and his masons go today. Schmidt, Hoepfel [stonecutters] and Dalles [blacksmith] are retained, the latter at $1 per day and found. The others are to be settled hereafter. At night patterns of hearthstones reviewed and worked on the models. Tymp stone, however, and wing stones postponed.

Three large blocks of sandstone resting on the bottom stone formed the back and sides of the lower hearth. The side stones (wing stones) had holes for carrying the nozzles of the tuyeres, which brought the hot blast into the furnace. The tymp stone, supported by cast-iron "bearers" set into the masonry of the walls, formed the major part of the front face of the hearth. The tymp stone did not go all the way down to the bottom stone, but was raised several inches above it. Below the tymp stone the molten metal was kept in the hearth by a dam stone, shaped like a prism, with a small aperture, the tap hole, cut at one side of its base. During smelting the tap hole was plugged with clay until the moment arrived for running off the iron into the pig molds. Both tymp and dam stones were covered with heavy cast-iron plates to prevent their being injured by heat and by the tools used by the founder or keeper in tapping the furnace or in pulling out slag. The space between the hearthstones and the rough walls of the furnace stack was filled with bricks or stones.[48]

Pauer takes over the journal:

December 4.—The wagons from Williamsport arrived in the afternoon [with the boilers]. Got them unloaded. [The round trip of 180 miles had taken 11 days.][49] The laborers were told that they would get only 87½ per day. Hoehn [carpenter] $1. Schmidt and Hoepfel complain of Dalles the blacksmith [that] he will not sharpen and steel their tools in the right way.

December 5.—Early in the morning Messrs. P. T. Tyson, J.H.A., and Mr. Oldfield left here. Rupprecht brought their trunks etc. to Cumberland.

Pauer was left with a greatly reduced staff. All the masons had gone, and fourteen other hands had been discharged. The mining force appears to have been left intact. During the rest of December, Pauer noted little activity except in digging at the sides of the furnace and dressing hearthstones. The stack, half finished, was covered with a plank shed to protect it from the weather, which became very bad.

During this same period ten wagon loads of machinery arrived. The steam engine and blowing machinery, which had been delayed in the hands of the builders in New York, reached Baltimore late in November. William Alexander had the various parts brought to his establishment near the wharves. Here he employed laborers to pack and load the machinery for transport to Lonaconing.[50]

On December 13 Ernest Wundsch,[51] the first company doctor, arrived.

As Christmas approached, Pauer noted in the journal:

December 23.—The Welshmen are drinking very hard. Steele thinks it cannot be helped.[52]

---

[46] The name of Granville Sharp Oldfield appears on Alexander's presumed list of stockholders. Oldfield, a native of England, had been a member of the New York merchant firm of Oldfield, Bernard & Co. until it was dissolved after a disagreement between the partners. Oldfield moved to Baltimore, where he engaged in business as a commission merchant and importer of wines. He was also the Baltimore agent for Lloyd's of London and for the New York countinghouse of Barclay and Livingston. Richard Wilson, the GCC&I Co. treasurer, had been Oldfield's bookkeeper in New York. Joseph A. Scoville, *The Old Merchants of New York City* (New York, 1862–1866) 1: pp. 75–76, 78–79. See also *Matchett's Baltimore Director* for 1837–1838 (Baltimore, 1837).

[47] Overman, *Manufacture*, p. 160; Overman, *Metallurgy*, p. 507; and Tomlinson, 2: pp. 77–78.

[48] Tomlinson, 2: pp. 77–78; Overman, *Metallurgy*, pp. 507–508; Overman, *Manufacture*, pp. 158–162; and H. R. Schubert, *History of the British Iron and Steel Industry* (London, 1957), p. 202.

[49] The distance is computed from point to point on the stage route, as set out in *Mitchell's Traveller's Guide Through the United States* (Philadelphia, 1836).

[50] See daybook entries, Nov. 28–Dec. 23, 1837, Welch and Alexander Record Books. The entries include drayage charges, laborers' wages, costs of wood and rope for packing, and the cost of weighing. One more load was dispatched in January and one in February 1838.

[51] Thomas and Williams (1: p. 537) report the name as "Wundreh." A careful study of the journals supports my reading, as does the manuscript census for 1840.

[52] ". . . the [Welsh] miners and mill hands 'had learned in Aberdare, Merthyr, Rhymney, &c., to drink beer like water, to get as drunk as tinkers, to swear and curse worse than demons of the bottomless pit.'" Rowland T. Berthoff, *British Immigrants in Industrial America 1790–1950* (New York, reissue 1968), p.146.

JANUARY 1–JUNE 30, 1838

Accustomed to the relatively mild climate of Baltimore and the Chesapeake Bay country, Alexander chafed at delays caused by the excesses of western Maryland weather.

As the working season of 1837 had been shortened by the hasty approach of winter, so that of 1838 was retarded by its slow retreat. Although we had hoped that the first of March would witness our commencement, the storms of snow, frequent in those mountains, continued through April, and it was not until May that the works were again fairly under weigh.[1]

However, it was Tyson, isolated at Lonaconing without intellectual companionship, who endured the tediousness of waiting for spring and the resumption of full operations. Daily life was monotonous and disorienting; he briefly betrayed himself by writing October and November instead of March in the journal entries. Nevertheless, he was able to push forward the development of the mines, the search for new beds of limestone (which was in short supply), and the completion of the "coal railroad" begun on September 20, 1837.

The coal railroad was a self-acting inclined plane 567 feet long with a 30 per cent slope. It connected the coal drift near the top of Dug Hill with the coke yard lower down near the top of the furnace. In 1838 it was also used to bring down building stone from the quarry.[2] Delayed by occasional shortages of laborers, carts, and sawed lumber for rails, the plane was finally finished on March 31, when trial loads of coal and stone were let down.

The cars ran on two parallel sets of tracks and were attached to a chain wound around a drum eight feet long and with a diameter of six feet located on the platform where the cars were loaded.[3] As a full car went down on one set of rails, its weight pulled up an empty car on the other. A brake on the drum wheel controlled the speed of the descent. The chain, of half-inch iron with links $1\frac{3}{4}$ inches $\times$ $2\frac{1}{2}$ inches on the inside,[4] was bought in Baltimore from Captain W. Graham, who had imported it from England. The cost was $272.26.[5]

The financial crisis of 1837 had passed, and the banks had resumed specie payments. With an easing of credit in England, British investors showed new interest in American securities. The George's Creek Coal and Iron Company sold all of its capital stock (much of it in London), and received full payment of $300,000 in the spring of 1838.[6]

Freedom from money worries permitted vigorous prosecution of the works, negotiation for the purchase of more land,[7] and location of a proposed turnpike between Lonaconing and the National Road. The company was also able to put under contract two projects not included in the 1836 estimates: a large sawmill and $7\frac{1}{2}$ miles of graded road extending from the sawmill yard to white pine forests at Beatty's Plains. The decision to build the mill was made because of the chronically short supply of sawed timber and the frustrating delays resulting from the scarcity of seasoned lumber and plank.[8] The several small sawmills in the vicinity, including one owned by the company, did not have the capacity to meet large-scale needs. During 1837 and 1838 the company had saw logs hauled to its own mill and got whatever small quantities of lumber it could from its neighbors. It also sent wagoners to Frostburg and Westernport to bring back dozens of loads of plank.

During the first half of 1838 the company gradually built up its working force, particularly of skilled miners. As the Lonaconing population grew, the journalists began to record difficulties with the workmen. Troubles ranged from drunkenness and quarreling to "extravagant" wage demands and strikes.

January 1.—Holiday. No work done. Mr. Tyson arrived late in the evening.

Tyson resumes journal:

January 2.—The ore drift has been carried in about 18 feet and [the opening] arched with stone part of the way. Some progress has been made at the open mining, having now about 200 tons out. Mr. Steele was directed to open a new drift in the bed of ore above the 8-foot coal. But little has been done with the coal mine, the men having refused to work at the price agreed on. Mr. Schmidt took two laborers and Mr. [Henry] Koontz and four additional horses to assist the wagon with the large cylinder [for the blowing apparatus] from the turnpike at Clary's [a tavern on the National Road 9 miles from Lonaconing]. He sends word tonight that after having started twice, they had only proceeded one mile. To-morrow they will get Clary's horses and try it again.[9] Carpenters putting up the drum wheel and brake at the head of the inclined plane.

January 3.—Mr. Schmidt succeeded in getting the cylinder to H. Koontz' [about 3 miles from Lonaconing].

[1] GCC&I Co. Report 1839, p. 6.
[2] *Ibid.*
[3] Journal, Nov. 3, 1837.
[4] *Ibid.*, Nov. 9, 1837.
[5] *Ibid.*, Nov. 9 and 10, 1837. See also W. Alexander to J. H. Alexander, Nov. 16, 1837, Alexander Papers; and daybook entry, Nov. 27, 1837, Welch and Alexander Record Books.
[6] J.H.A.'s statement of 1850, Alexander Papers. A partial list of the English stockholders includes the London banking

firms of Smith, Payne & Smiths and Magniac Jardine & Co., various members of the Smith family, and William Smee, accountant-general to the Bank of England. Magniac Jardine to J. H. Alexander, Oct. 18, 1845, John H. Alexander Collection, Maryland Room, University of Maryland.
[7] Its principal acquisition was a 500-acre tract called "Buck Lodge," with a frontage of about $\frac{3}{4}$ of a mile on the Potomac River. At this site the company expected to build wharves and construct a basin which would in effect extend navigation from the end of the C&O canal up river about a mile above Cumberland. GCC&I Co. Report 1839, p. 8; and J. H. Alexander to J. P. Ingle, Sept. 15, 1838, C&O Canal Records, National Archives.
[8] GCC&I Co. Report 1839, p. 7.
[9] The cylinder was 8 feet long and 5 feet in diameter. Walter R. Johnson, *Notes on the Use of Anthracite in the Manufacture of Iron* (Boston, 1841), note, p. 8.

January 4.—The wagon arrives at 11 A.M. with the blowing cylinder all safe. Nodules [of iron ore] begin to appear at the opening of the new drift (No. 2). Uncovering of limestone at Totten's going on.

January 5.—Directed Mr. Steele to construct a temporary tramroad from the ore mine towards the furnace. As the drift No. 1 proceeds forward, the ore No. 15 becomes much more abundant. It now consists of closely compacted nodules which nearly fill the whole space between the two small beds of coal.

The 1840 report of the Maryland state geologist includes a plate showing the ore beds worked at Lonaconing. Both the geologist's report and the plates of the company's 1836 "memoir" show clearly the nature of the deposits in Dug Hill. Beds of iron ore were sandwiched between beds of shale, sandstone, clay, and coal. In this locality the veins of iron within an ore bed ranged in thickness from a few inches to several feet.

January 9.—Mr. Schmidt's men were directed to uncover a portion of [another bed of limestone] where it crops out in the northeast bank of Koontz Run. There also appears at this place above the limestone a fine grain sandstone, which splits easily into parallel layers and will be valuable for flagging and for the inwalls of the furnace, etc. Wrote to J. Harris directing [him] to come here himself on the 20th February and to have 20 masons more on the 1st March.

January 10.—One car has been put on the railroad from the [iron] mine, and they have commenced placing the ore on the leveled space between the mines and the furnace. The two last stones (it is hoped) for the hearth were hauled this afternoon.

January 13.—P. T. Tyson preparing to leave for Baltimore and has left the necessary directions for the prosecution of the various kinds of work.

Pauer takes over the journal:

January 15.—The miners did not go to work, as Mr. Steele says, in consequence of a frolic on Saturday. None of them, however, could get liquor on the premises; so all was quiet.

January 18.—F. Pauer made a bargain with Godfrey Miller [local farmer] to build two cabins, the same dimensions as those on Dug Hill, which he built, at $40. Hauling done by the company. He has to commence tomorrow. Mr. Steele discharged William Williams, one of the miners.

January 19.—Steele gave Williams work again after he promised to keep sober. The blacksmith Huhnlein went to work again. He works in the daytime, the other one by night.

January 23.—In the afternoon Steele showed F. Pauer two pretty stout springs in the [mine] shaft. Arner had to make a trough to lead the water to one side. Steele thinks he can go on in the same shaft.

January 25.—Moses Ayres bought a horse for the use of the railroad and to carry the mail. [The Ayres family was one of the first to settle in the George's Creek valley.] The stonecutters did begin at laying the hearthstones.

January 26.—The stonecutters found that the stone at the bottom of the hearth was not laid level. They had to dress it.

January 27.—Heavy snow storm. The hands above the ground had to quit at 9 o'clock. The stonecutters could not lay the hearthstone, as the wind did drift the

snow in the furnace. Schmidt, a German blacksmith, arrived from Baltimore.

January 29.—[August] Weisskettle and [John] Wortman, two masons, arrived from Little York, Pennsylvania.

January 31.—Steele discharged three of the miners, as they got drunk.

February 1.—Steele gave an account of the two contracts. The open work amounted to 233 cubic yards, $233; the drift to 14 yards, $77. [Both coal and iron mining were on a contractual basis, the men agreeing to a price by measurement or by weight.]

February 2.—A new contract for the remaining part of the open work was made at 75 cents per yard instead of $1.00. For the underground work, some allowance for water had to be made—$5. Arnold, the blacksmith, was discharged.

February 3.—As it was found out by F. Pauer that the miners who got intoxicated bought the liquor at [William] Koenig's shantee, the boarding house of Koenig was broken up. He was not discharged, as he was not at home when the liquor was sold.

February 5.—F. Pauer settled with the working people for the month of January.

February 6.—Joseph Shriver [cashier of the Cumberland bank] had not received a draft from Baltimore and could consequently not pay the checks [for employees' wages.]

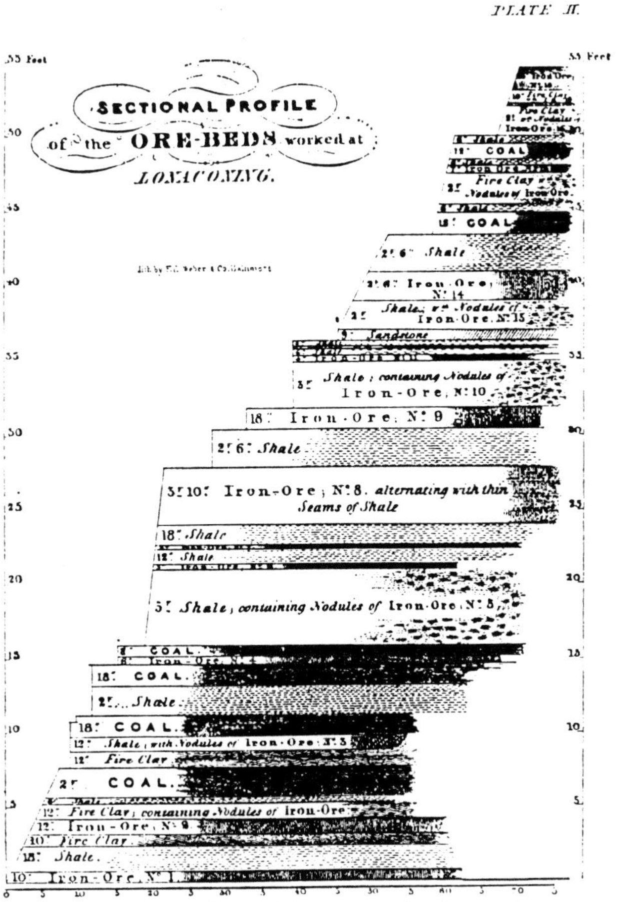

FIG. 5. Lonaconing ore beds. *Annual Report of the Geologist of Maryland, 1840.*

February 10.—F.P. went to William Shaw to see whether he would now be able to attend to the business about the turnpike to Clary's.

On January 20, 1838, the Maryland General Assembly passed an act authorizing the George's Creek Coal and Iron Company to construct a "Turnpike Road" from Lonaconing to intersect the National Road at Frostburg or at some point between Frostburg and Cumberland. This would be a toll road with rates not higher than those charged on the National Road. Allegany County delegates told Thomas Alexander that there would be no local opposition to the project and said that landowners would probably give the company the right of way in exchange for permission to use the road free of charge.[10]

February 12.—Schmidt, the new blacksmith, got sick. Dr. Wundsch complains that the people want so much medicine that he has lost money. Last month he had to give them $29 worth, and this month, up to the 12th, $23. F.P. advised him not to ask the monthly payment, but to make out monthly bills for those having consulted him.[11]

Tyson resumes journal:

February 15.—P.T.T. arrives from Baltimore in the evening. About 6 inches [of] snow fell last night. It commenced snowing again in the afternoon, which was succeeded by hail which continued most of the night.

February 16.—The hail continued to fall until about 10 A.M., and then after a few hours it began to snow with high winds from the westward, which continues late in the evening. Weather very cold. There appears to be about 450 tons of ore out. The drift has been pushed about 40 yards. Not much has been done with the coal mine. The brake [for the inclined plane] is not finished for want of the iron work. The rails on the inclined plane not laid for want of the sawed stuff. No progress has been made in framing the molding house. A considerable quantity of limestone has been uncovered on Koontz Run. Two more hearthstones beside the tymp [stone] are to be dressed.

February 17.—Thermometer at 9 A.M. stands at 4°. But few hands worked until after dinner in consequence of the severe cold. Low water prevents the sawing of railroad stuff, which is much wanted. Two of the carpenters have been for some time engaged in making trams and wheelbarrows.

February 19.—One new hand at the mines today, but not a full [experienced] miner. Drift No. 1 going on night and day, and proceeds nearly a yard forward in 24 hours. It is 6 feet high and six feet wide. Three hands are beginning to go underground with Drift No. 2 and are widening a portion of the open entrance preparatory to walling and arching, for which two masons are preparing the stone. Two colliers commence the coal drift by the day. Some hands are blowing out [blasting] foundations for the walls on each side of the furnace. Contracted with a mountaineer to build an ice house and fill it.

February 21.—Mercury stands at −1° at 7 A.M. The mornings have been so cold for several days past that the stonecutters at the hearth could only work in the afternoons.

February 22.—Thermometer at 7 A.M. stands at −9°. It is so cold that a number of hands could not work out of doors. Carpenters again stopped working at the molding house in consequence of the severe cold.

February 23.—John Thomas, who has been here since October last and was intended for a furnace keeper, will persist in getting drunk and has been discharged.

February 25, Sunday.—Mercury down into the bulb early in the morning, so that temperature is below −14°. William Shaw reports progress and is directed to procure right of way for a railroad to Westernport. Coleman with the last of the steam engine gets within 5 miles of the place, and the wagon slipped off the road, but did not upset. It will, however, be necessary to unload and make use of sleds to bring the load down.

February 26.—But few of the miners at work today. The water in Drift No. 1 is troublesome. Two masons are walling and arching the mouth of Drift No. 2. Three masons commence the wing wall on the northeast side of the furnace. Two stonecutters dressing tymp bearer and one splitting out tymp stone. Mr. Schmidt goes with the horses and some hands to unload the wagon and bring the matters on it down in sleds. He gets the cast-iron pipes here and will bring the balance on tomorrow.

February 27.—Ore No. 19 comes in very thick and rich, or at least heavy. Hopkins [the founder] delighted with it. Mr. Schmidt succeeds in getting the balance of the wagon load and the wagon itself here. Blacksmiths finish the iron work for drum and brake [for inclined plane]. The connections for the chain remain to be finished. Mr. Hopkins sets a clamp with a ton of iron ore No. 15 and fires it.

Before iron ore could be used in the furnace, it had to be roasted to get rid of undesirable substances (principally sulphur and phosphorus) which would make for low-grade pigs. At Lonaconing the roasting was done in clamps similar to those used for limestone and coke. The workmen first broke the ore into pieces about three or four inches in diameter. Then they built the clamp, which would generally be about ten feet wide, twenty or thirty feet long, and ten feet high. The bottom layer was coal, arranged so as to provide a draft. Next came alternate layers of iron ore and coal until the clamp reached the desired height. A rough rule of thumb was one inch of fuel for each foot of ore, the ore layers being usually eighteen to twenty-four inches thick. At Lonaconing the proportions were determined by weight. Small clamps used one hundred pounds of coal for each thousand pounds of ore. The whole clamp was covered with small ore and then set on fire at the bottom, "in the direction of the prevailing wind."[12]

February 28.—The miners at the drifts stopped work and demand higher wages. They get $6 per yard forward

---

[10] Md. *Laws* Dec. Sess. 1837, ch. 11; and T. S. Alexander to J. H. Alexander, Dec. 30, 1837, Alexander Papers.

[11] For 50 cents a month withheld from wages, the hands were entitled to medical services for themselves and their families. "Rules of the Lonaconing Residency," preceding journal entry for Sept. 21, 1838.

[12] J. H. Alexander, *Report on the Manufacture of Iron* (Annapolis, 1840), pp. 144–145. See also Frederick Overman, *The Manufacture of Iron in all its Various Branches* (Philadelphia, 1850), pp. 43–45, and *Treatise on Metallurgy* (6th ed., New York, 1882), pp. 289–290.

and have earned about $1.90 per day each. Their demand was therefore resisted. Towards evening the writer is informed they give some indications of reconsidering the matter.

March 1.—The two Welsh colliers at the coal drift will not accept $3 per yard forward, and the contract is given to a Frostburg miner at that price. The upper bench containing ores from No. 9 to 12 inclusive has been let out to 5 of the miners at $1.25 per ton for the ore, and they find their own powder. This bench is 6 feet in height, and each square yard is found to produce a little more than a ton of ore. The drifters [workers at the drift] are still hanging on, and will probably go to work again upon being allowed something whilst so much water comes upon them from the strata in the top of the work. The amount of ore obtained up to the 28th February 1838 and stacked up is 269 cubic yards. [The journal here contains a list of the amounts produced from each of the workings.]

March 3.—P.T.T. goes to Cumberland to take a draft to the bank for the February payments. He meets with a Welsh miner, and directs him to appear at Lonaconing.

March 5.—Snowed all night and continues most of the morning. The depth of the snow was 10 inches, but the weather becomes warm, and the snow begins to melt. Laborers and miners do not work. The drifters, it is believed, will go to work again. Another miner will join them who has arrived. Mr. Hopkins is packed off to Frostburg on his road to Pottsville [Pa.] for miners.

March 6.—The drift No. 1 is recommenced today with 4 hands. It is difficult to keep the carpenters at work that is wanted for want of lumber. Blacksmiths making weighing machine, and making and sharpening tools. Weather getting warmer, and the walking execrable.

March 7.—As warm rains may now be looked for at this season, there is some reason to fear an inundation at Lonaconing, particularly as the natives say there is more ice in the creek than has ever been observed before. Directions have been given today to cut away the trees on the margin of the creek from the carpenters' shop upwards, about 100 yards, for the purpose of allowing the ice to pass off when the freshet shall come. The tymp stone is brought as far as Steele's house and the sled gives out, but is again repaired, and it is expected the stone will be brought to the furnace on tomorrow. Directed an additional cabin to be erected near the graveyard. [James] Rowles putting timbers in the coal drift.

Both ore and coal drifts had to be timbered to prevent the earth from slipping in. The job was usually done by skilled miners. On each side of the drift they put in six-foot posts about a yard apart, resting them on solid rock to prevent their sinking. The posts were not vertical, but slanted slightly inward. Across the drift, resting on top of each pair of posts, was a cap of heavy timber ten or twelve inches square and about five feet long. Split timbers about two or three inches thick covered the caps and posts to prevent earth and gravel from falling into the drift. The timber used in ore mines was preferably locust, white oak, or red oak. In coal mines, pitch pine was more durable.[13]

March 9.—Weather more moderate still, and *no snow* falling, but what is on the ground is become very soft and makes pedestrian locomotion very unpleasant. The tymp stone is hauled today.

March 10.—Drift No. 1 is impeded by water, but is going on with 2 sets of hands 8 hours each. It will be driven on night and day as soon as two more drifters can be got. One additional has arrived from the Cacapon tunnel today.[14] Arching of drift No. 2 finished *at last*, and the drifting will proceed as soon as miners can be had. The frost coming out of the bank on each side of the furnace, and the bank falling down considerably. Ice cleared from about the sawmill, but there is not water enough to saw.

March 12.—Mining going on well except that the ore cannot be hauled away fast enough for want of cars, one being broken. Another miner has come from the canal. Blacksmiths making the connection for the inclined plane chain. Two additional double cabins have been contracted for.

March 13.—Blacksmiths at weighing machine and tools, the iron work for the brake being finished, and the chain hauled to the top of the plane. Letter to J. Harris, mason, ordering him to make his appearance forthwith at the head of his gang.

March 14.—Coal mine going on with one hand, but another has been engaged. Three masons got at the wing wall after the fallen earth and stones were removed. About 40 or 50 cart loads fell down last night.

March 15.—The ice in the creek breaks up and lodges opposite Koenig's house to such an extent as to threaten an inundation of the town. It was attacked and cut to pieces by a party of Germans, with a loss on their side of one axe.

March 17.—During last night the falling weather changed from rain to snow, which has continued all day to fall very fast. No work could be done out of doors except to shovel the snow off the shed over the furnace and the bank behind it. This [has] been twice done. The shed at the front arch of the stack came down. Mr. Harris, the mason, arrived today and brought a miller with him. Mr. H. has 8 masons and a hod carrier on the way. The snow at 9 P.M. is 18 inches deep.

March 18, Sunday.—Snowing continued all night and until 2 P.M. today. The whole thickness 2½ feet. A number of hands all day clearing about the furnace. No mail today or yesterday. Rupert is directed to try to take the mail to Frostburg early in the morning.

March 19.—The laborers have been shoveling snow all day. The men at open mining get the snow out of the way by dinner time and go on with their work. A land slip has taken place into the ore road, covering some of the ore. The extent of surface which came down was 35 feet by 20 feet. It contained a tree of considerable size, which maintains its upright position. Seventeen men sent by Walters [from Baltimore] to work at open mining, and one drifter arrived today. Mr. Harris' 8 masons and 1 hod carrier arrived, and there was considerable difficulty in getting them housed for the night. Rupert succeeded in transporting the mail to Frostburg and back with great difficulty.

March 20.—Laborers employed in clearing away the dirt which has fallen from the bank above the ore road and at the sides of the furnace.

March 21.—Part of the new Irish hands commence uncovering ore beds in the rear of the front row. The American newcomers join in with Walters at the bench containing ores from 9 to 12. The German laborers

---

13 Overman, *Manufacture*, pp. 61–64.

14 The Cacapon tunnel, now known as the Paw Paw tunnel, carried the Chesapeake and Ohio canal through the mountains 30 miles east of Cumberland. The tunnel contractor sent an agent to England to secure practical miners to work on it. Chesapeake and Ohio Canal Company, *Ninth Annual Report* (Washington, 1837), p. 8.

tending masons and clearing away the fallen earth near the furnace. Divers accidents occur today. Conrad Blatter cuts his foot with an axe. A boy is cut *a posteriori* with an axe. The wagoner mashes his finger. A tram car tumbles from the road down to the steam engine plot. An avalanche of earth and stones comes down upon the northeast wing wall, which the masons had just quitted for dinner.

March 22.—The wing wall 13 feet high. A man and a boy got hurt today in handling stones.

March 23.—Four of the Germans assisting the masons strike for higher wages, but go to work again. The weather has been very warm, and the melting of the snow on the north exposures of the adjoining mountains caused a great freshet in the creek in the afternoon. The water overflowed near the shantee of Zacharias. A dam of brush and stones was made across the low place. Some brush and trees which had been lodged in the channel were gotten out, so that by sundown the creek was nearly confined to its banks. Rupert goes to Frostburg with the carryall for the masons' baggage, but in consequence of the high water, leaves the concern [vehicle] at H. Koontz' house. He reports the arrival at Frostburg of a miner sent on from Pottsville by Mr. Hopkins, who says Mr. H. will arrive tomorrow if he is not disappointed in a seat on the stage.

March 24.—Mr. Hopkins arrives with 4 miners, having left two at Hagerstown to walk on, the stage being full. [It was 74 miles by stage route from Hagerstown to Frostburg, and about 8 miles from Frostburg to Lonaconing.]

March 25, Sunday.—The two remaining Welsh miners arrive.

March 26.—Gave Harris the directions for the chimneys from the hot air furnaces [to heat the blast], which chimneys will be carried up in the wing walls.

March 27.—A fracas occurred last night [at Moses Ayres's house] between some of the Baltimore County men and a portion of the Welshmen engaged at drift No. 1. Orders were issued that whoever shall attempt to provoke a quarrel hereafter shall be discharged. One of the drifters (of No. 1) starts for the [Cacapon] tunnel for his family, and most of the rest go to Frostburg to make some purchases for housekeeping. It is doubtful if they will work again this week. Drift No. 2 is commenced to be taken underground today.

March 28.—The southwest side of the furnace cleared of the fallen earth, and two hands are blasting out the limestone for a good foundation [for the second wing wall]. A load of goods arrives from Baltimore, including five sets [of] railroad wheels.

March 30.—Masons commenced the southwest wing wall. A tramroad to the foot of the quarry commenced. A route for a tramroad to the limestone quarry located by Mr. Pauer.

March 31.—One load of coal and one of stones let down the inclined plane. It is a good fixture. Amount of ore obtained during this month [listed by ore bed numbers] 105 tons. Drift No. 1 is now advanced 59 yards (10 yards this month). The drift into the coal has been carried in 16 yards this month and is now 21 yards from the mouth.

April 1, Sunday.—It was proposed to the Irish to go on with the uncovering at open work No. 2 at 15 cents per cubic yard or clear out [leave]. They were disposed to kick up a disturbance, but have got quiet and will go off tomorrow. Being much dissatisfied with the results of the open mining No. 1 from the ignorance and laziness of the hands, means are being taken to reform that department or discharge the whole pack of Baltimore County men. They have undertaken to remove a bench [at 40 cents a cubic yard] as before stated [journal, March 31],

but on trial for one week with Mr. Steele standing by all day long. Mr. Hopkins to attend to the rest of Mr. Steele's duties.

April 2.—The covering taken off the stack in order to commence laying stone. The quarry road (tram) nearly all laid, and will be put in use on Wednesday. The Irish are sent off, and also part of the Baltimore County men, the best of whom are retained. One of the Pottsville Welshmen starts for Pottsville to bring the families of the rest and his own. P.T.T., having arranged everything to suit, proposes to leave for Baltimore on tomorrow.

Pauer takes over the journal:

April 4.—F. Pauer gave to J. Harris the levels of the four walls of the stack. Two of the empty cars on the railroad ran off. One man got hurted.

April 5.—Harris wants 5 tenders more and some hands. Altogether he thinks 9 more will do. The bridge at the mill broke down. One of our horses had pretty near been killed. Another car ran off from the rail track today.

April 6.—The hauling on the railroad shows a great deal of difficulty. The iron for the bolts is of an inferior quality, and thereby the cars often times run off.

April 7.—F. Pauer sent for 9 hands to Cumberland to come here by Monday morning.

April 9.—The masons were started at the stack. Another driver was engaged at $15 per month and found. An addition to the store was commenced. Arner is engaged at fixing the center pole in the stack. [They were about to set up the tram for working on the curved inner walls again.]

April 10.—Instead of 9 hands, 12 arrived. Mr Harris is of [the] opinion that they can all be employed to advantage. Arner engaged a carpenter. He says he must have a hand more to get on. Yesterday Henry Koontz returned from Wheeling and brought 1,700 pounds of bacon at 9½ cents per pound. John Thomas [keeper] was discharged by Mr. Hopkins.

April 12.—Another cabin up on Dug Hill for the coal miners was ordered to be built, and the last one in the flat got finished.

April 13.—Several of the hands did not work, today being Good Friday. Another car broke loose from the railroad. Steele says it is owing to the bad quality of the iron, but Mr. Harris says it is the carelessness of the men attending to it.

April 15, Easter Sunday.—Heavy snow storm all the day.

April 16.—Most all hands not at work. The snow lays from 7 to 8 inches deep. Six hands were put to work half a day at clearing the snow from behind the furnace.

April 17.—Last night at a party at Herman Dittmer's shantee the masons had a quarrel with some of the German hands. F. Pauer consulted with Mr. Harris. Two of the tenders were turned from the stack, and some of the masons got a warning from Mr. Harris.

April 18.—Mr. Hopkins wants another cabin for two families from Pottsville. F. Pauer hired [Charles] Broadwater [farmer] to do it. Four dollars in addition were allowed for hauling the logs.

April 19.—A new miner from Pottsville was engaged by Mr. Steele.

Tyson resumes journal:

April 21.—Mr. Tyson arrived. The fixtures at the inclined plane work well, and the stone is supplied to the masons by it as fast as they can use them. The stack has been carried to 26 feet in height. Drift No. 1 and the left-hand heading therefrom have progressed well for the last 2 weeks. Drift No. 2 also goes on and yields ore in

great abundance. The coal drift progressing, and a heading is going off from it.

Underground mining began with a drift or tunnel about six feet high and four feet wide driven into the side of a hill where a workable bed of ore or coal existed. As soon as the drift was carried in far enough, the miners made side openings (headings) into the vein at right angles to the drift and opened up stalls or rooms for taking out the minerals.[15]

April 24.—Mining going on as before except that 2 Welshmen are engaged at undermining at the open work by taking out ore No. 4 with the coal above and below it.

Iron ore No. 4 was a six-inch band lying between eighteen inches of coal below and eight inches of coal above. Five feet of shale formed a roof over the top coal; two feet of shale lay beneath the bottom coal. Instead of removing the overburden of shale, the miners cut under it along the side of the hill, using props to support it as they went in and removed both coal and ore.

April 25.—Some of the people have seemed prejudiced against Mr. Steele for a long time, and have tried to get him away. They now accuse him of feeding his mare with other people's oats. Mr. Steele insists that the charge is not true, and that as he is satisfied of a determination on the part of a number of people here to drive him away, he can have no peace with them, and he will therefore leave us. He has settled up his accounts etc. and proposes to go off tomorrow.

April 26.—In consequence of Mr. Steele's going away, there has been much to do in arranging the mining and placing it under the charge of Mr. Hopkins. The main drift No. 1 is infested with carbonic acid so that the lights will not burn at the fore-end, which is carried in the coal above No. 19, leaving the ore in the bottom. [The drift is now being driven forward with the coal as its roof.] Order the men to take out this ore and the fire clay below it so as to give a height of 7 or 8 feet in the drift [and allow for more ventilation]. In the side heading the air seems more pure. It does not produce much inconvenience if the weather be dry and windy.

The carbonic acid referred to above was CO_2, known to coal miners as "black damp." This non-explosive gas results from the absorption of the oxygen in the air by the coal in the mine and the concurrent formation of carbon dioxide. It is also produced by normal activities underground: burning lamps and candles, firing explosives, even the breathing of the miners and their mules or ponies. If there is enough black damp present, it extinguishes lights and causes death by suffocation. Black damp was most troublesome on still, rainy, days. In the 1830's the only way of ventilating a mine was to drive in an additional heading and connect it with an air shaft.[16]

April 28.—Warm day. After 3 P.M., a tremendous storm of wind, hail, rain, thunder, etc. came over and stopped work. Miners continue to apply to us faster than we can give places to them. The families of the Pottsville miners arrive.

May 1.—The force of miners has now increased to 36 besides those engaged in the open work. Eight of them are to proceed with the coal openings, of whom 4 will drive in the main drift, and 4 the lateral heading to the left.

May 3.—The stack is now leveled off at 30 feet above the hearth, and two of the fourth set of binders put on.

May 4.—Settling up accounts and paying off men for April. [Wrote to] Joseph Shriver and enclosed a draft for $2,000 and one for $1,500 for deposit in bank [at Cumberland].

May 5.—Rain continues until 9 o'clock, when it becomes mixed with snow. Snow continues to fall for 4 or 5 hours steadily, and afterwards the wind changes to northwest with squalls of snow the whole evening.

May 7.—Drift No. 3 proceeds, and is just about to go underground. Cabin builders going on with their work as usual and also building small houses near the drifts for tools.

The number of tools used by fifty miners represented a considerable amount of money and labor. The company supplied each miner with a pick, a mallet for driving wedges and drills, a wedge (driven into crevices to split off pieces of rock or ore), a five- or six-pound sledge hammer for breaking larger pieces of rock and ore, and a pointed shovel. All of these had to be kept in good repair by the blacksmiths. In addition, some of the men had blasting tools: hand drills, scrapers, tamping bars, and priming needles.[17]

May 8.—Weather cold, occasionally spits of snow. The hauling of sand and lime is difficult on account of the muddiness of the roads. The bank on the northeast side of the furnace being saturated with water has moved down somewhat and done some injury to the unfinished wing wall. Ordered Schmidt, the blacksmith, to be discharged for divers reasons.

May 10.—Mr. Farron and his brother arrive from Mr. Kemble [West Point Foundry] to put up the engine.

May 11.—There has been a good deal of blasting in drift No. 1 for some days, and the air has become very foul. The airway is to be driven on as fast as possible. Tram rails are being laid as they are required in the drifts.

May 15.—Purchased two horses and gear [harness] from Moses Ayres for $150. Some petty thievery has recently taken place. A watchman has been engaged to patrol the town every night.

May 16.—The upper part of the stack being thinner than below requires a greater proportion of face stones than are obtained from the quarry, and the masonry does not go on as rapidly as it might. More hands have been set to work at the quarry. The excavation for the foundation of the steam engine goes on. Messrs. Farron at work arranging the pieces [of the engine]. The two blacksmiths, Schmidt and Huhnlein, decamp without notice and leave us in a sad plight. A mountaineer is engaged as a makeshift.

May 17.—The stack was leveled for the 5th pair of binders (36 feet 6 inches).

May 18.—Wrote to R. Wilson enclosing an advertisement for brickmakers.[18]

[15] Overman, *Manufacture*, pp. 64–67.

[16] See Sir Richard Studdert Redmayne on "choke damp," *Encyclopaedia Britannica* (14th ed.). For black damp problems in Maryland coal mines, see Katherine A. Harvey, *The Best-Dressed Miners* (Ithaca, 1969), pp. 34, 43–47.

[17] Overman, *Manufacture*, pp. 55–56.

[18] In the Baltimore *American* of May 24, 1838, the company advertised for a brickmaker to produce 500,000 or 1,000,000

May 19.—Messrs. Swartwout, Bruce, Howell, and Mc-Culloh visited us today.

These were the first of a succession of interested visitors during the year. Upton Bruce was one of the incorporators of the Allegany Iron Company, which by 1828 had already erected a furnace and forge in what is now Garrett County, Maryland.[19] Lewis, Benjamin, and Henry Howell were the incorporators of the Maryland and New York Iron and Coal Company, which later built iron works at Mount Savage, and George McCulloh was one of the incorporators of the Maryland Mining Company, one of the largest coal producers of the area during the 1840's.

Samuel Swartwout was collector of customs at the port of New York, a post given to him as his share of the spoils after Jackson's election in 1828. It later appeared that Swartwout had appropriated for his own use more than a million dollars in public funds, some of which he had invested in Maryland coal and iron lands. His term of office expired in March 1838. In August 1838, before his embezzlement was discovered, he fled to England, hoping to dispose of some of these lands which he had bought from Matthew St. Clair Clark, president of the Maryland Mining Company, and from the Howells, incorporators of the Maryland and New York Iron and Coal Company, in which he owned over 1,400 shares.[20]

Alexander takes over the journal:

May 21.—J.H.A. (who arrived sick in Frostburg on Saturday evening) reached here today. Mr. [Alfred] Duvall, the millwright, arrives, who makes an estimate for the sawmill with two frames to cut 5,000 feet per day and 40 feet long. [Here Alexander itemizes the estimate: Duvall's own work, $1,900; materials (principally castings and wrought iron), $1,616; fares and transportation of seven hands, $154; transportation of materials, $300; hewing timbers, $200. The whole mill would cost about $5,000.] Mr. Farron cuts some didos, so that he is rather uneasy in his employment.

May 22.—J.H.A. and P.T.T. reconnoitered for the extension of an avenue down to the Totten lot [south of Koontz Run] the proposed site of the village. Stack for the first time supplied this afternoon with face stones. Determined to discharge John Farron for drunkenness and misbehavior.

May 23.—John Farron discharged this morning, but upon contrition and his solemn promise of amendment was

bricks at the Lonaconing site. The contractor would be required to have at least "two sets of hands," and he must have 40,000 or 50,000 bricks ready for use within a month after signing the contract. The bricks would be inspected by "the master brick layer," and would be paid for upon delivery. The advertisement was also carried by newspapers at Chambersburg, Carlisle, and Gettysburg, Pa.

[19] Md. Laws 1827, ch. 197.
[20] Dict. Amer. Biog.; Joseph A. Scoville, The Old Merchants of New York City (New York, 1862–1866) 2: pp. 251–261; Benjamin Silliman, Extracts from a Report Made to the New York and Maryland Coal & Iron Company [sic] (London, 1838), pp. 17, 46, and 48; Cumberland [Md.] Alleganian, Oct. 16, 1847; and Allegany county land records, liber W, folio 175.

permitted to come back. Various miners appear at the door with loud complaint against the boss, Methusalem Rowland, which was referred to Hopkins for adjudication. The back, or retaining wall, will be used now as one of the abutments of a stone arch instead of a wooden bridge as was contemplated before. [This is the bridge for the top house. The journal here contains a technical discussion of the construction of the arch, including sketches and calculations.]

May 25.—At night agreed with Mr. Hopkins for a permanent salary, he sending for his family. Terms: the expenses over [passage money from Wales] of his family to be paid by us, and his salary to be rated at $1,300 per annum from the 1st June proximo and to be paid up in full when the furnace is in successful blast. The wagon arrives from Baltimore, and the patent safety fuse [is] tried with tolerable success.

Safety fuse was introduced in the Cornish mines early in the nineteenth century. In this country it was used in construction of the Erie Canal (1818–1825). In the 1830's it was advertised for use in quarries, and by 1839 was said to be generally used in the Pottsville, Pa., mines. It sold for one cent a foot. The fuse was described as resembling varnished cord about $\frac{3}{16}$ of an inch in diameter. It burned slowly (about 18 inches a minute) and allowed the miners plenty of time to get out of the way before the charge exploded.[21]

May 27, Sunday.—Experiments with the thermobarometer gave us as from notebook 208.65° of the instrument as the point of boiling water.

The journal here contains calculations relating to the experiments. From these experiments Alexander deduced that one degree difference in boiling point would represent an altitude difference of roughly five hundred feet. A preoccupation with reading barometers and thermometers, beginning with the journal entry of May 22, was connected with plans for surveys of possible routes for roads and railroads for marketing the company's products. In mapping the rugged terrain, the engineers used the technique of leveling, but they could also use the barometer to obtain heights more rapidly and more simply. Furthermore, the use of the barometer gave them a rough check on their surveying instruments.

Tyson resumes journal:

May 30.—J.H.A. and the writer leave for Cumberland at noon, the work going on well. Negotiations going on with the Shrivers for land for a coal depot above Cumberland [Buck Lodge].

June 1.—P.T.T. leaves J.H.A. at Cumberland and arrives in Lonaconing at 11 A.M. J.H.A. intends to set out for Baltimore. At 10 A.M. John Welsh, one of the men at open mining, is blown up by his own carelessness and instantly killed. The miners out of respect to the deceased do not work the balance of the day.

June 2.—The excavation for the foundation of the steam engine masonry is finished, and the drawings therefor

[21] See Portland [Me.] Advertiser, July 27, 1838; Pottsville Miners' Journal, March 23 and April 6, 1839; and Jour. Franklin Institute, May 1839: p. 358.

Fig. 6. Mining iron ore. Overman, *Treatise on Metallurgy*.

handed to Mr. Harris. The back part of the stack is up to the point where the top house arch will start. The front [is] still higher. The miners do not work and attend the funeral of John Welsh at 2 P.M. He was interred at Mr. Shaw's chapel.[22]

June 4.—The masons commence the wall upon which the steam engine will rest.

June 6.—The Farrons are discharged for having been drunk for the last 3 days.

June 7.—Mr. Graham arrives at Frostburg.[23]

June 8.—The state of the store and accounts etc. is explained to Mr. G. The ore in drift No. 2 suddenly appears of the extraordinary thickness of $4\frac{1}{2}$ feet! entirely across the drift. Mr. Harris has contracted with a new man to put up the powder magazine at 75 cents per perch. It becomes necessary to fix new prices for the drifts from changes in the ground as we go in. The miners hold out for extravagant prices, but are refused, and they do not work.

June 9.—The drifters still idle, but in the afternoon No. 1 give in and agree to take $6 per yard forward. The front of the stack is 50 feet high, and the masons cannot do more work on it until the back [retaining] wall is carried up to the level of the arch which will connect it with the furnace.

June 11.—In the afternoon the drifters in No. 2 agree to work for $16 per yard forward, the ore being 5 feet thick and very hard. Drifters [in] No. 3 give in also, and contract for $8 per yard. A contract made with two miners for timbering the mouth of the coal mine at $1.50 per sett (or pair of timbers, as the Welshmen call the three making a sett) [that is, two posts and a cap, as described in the text following the entry of March 7.]

June 12.—Ledley and his brickmakers arrive.

June 13.—Brickmakers preparing the yard, and the shed being put up.

June 14.—Brickmakers commence molding.

June 16.—The airway from drift No. 1 has reached daylight about half way between No. 1 and No. 2 drifts. The undermining at open work having been completed far enough in, the props are knocked out and a portion of the front of the superincumbent beds falls down and shows a good deal of ore. Plank is scarce. Everybody who has promised to send it disappoints us.

June 18.—Letter from Alfred Duvall, with list of lumber wanted for sawmill, which is copied below. [This is a detailed specification of 360 pieces of white oak and poplar needed for the mill wheel, together with an order for 4,000 feet of two-inch white oak, poplar, or chestnut plank for the forebay and trunk which would supply water directly to the wheel.] Weather hot and has been so for 3 weeks past.

June 19.—A Mr. [H.J.] Welch and 4 hands arrive [from Baltimore] to do carpenter work by the day, and commence at the stone houses. A wagon load of round iron arrives from Oliphant, and a load of flour from Stockton & Byers. Ordered a load of bar iron from W. Bowers [owner of Warren furnace and forge in the southwestern corner of Franklin County, Pa.].[24]

June 20.—The masonry of the back wall is carried to the proper height for starting the arch to the stack [for top house bridge], and part of the masons work at the southwest wing wall. At open work a large portion of the undermined beds fell down, to the great hazard of the miners, but no one was hurt. Mr. Alexander arrives.

Alexander takes over the journal:

June 21.—In the morning P.T.T. and J.H.A. explore on the valley of Koontz Run the location of a supposed bed of limestone which seems to present considerable facilities for a supply. Reconnoitered also the route of a proposed common road leading along said valley ultimately for the advantages of the white pine forests on Beatty's Plains and collaterally to be used for the limestone if upon further examination it shall seem expedient, and also for certain beds of building stone. At the furnace the centers for the bridge house arch are put in. The masons at the steam engine wall are assisted very much by the arrival of the new engineer, Yonger, who comes today with Mr. Graham. In the afternoon definitive construction of patterns for the ring stones [arch stones] for the bridge house arch. [The journal here contains a sketch and calculations.]

[22] William Shaw's father, one of the early settlers in the George's Creek valley, was the first Methodist local preacher in the region. He was ordained as an elder by Bishop Francis Asbury in 1813. Shaw held regular services at the small meeting house he built at Shaw's Mill (now known as Moscow), about four miles southwest of Lonaconing. In 1838 traveling Methodist clergymen held occasional services here. Francis Asbury, *Journal and Letters*, Elmer T. Clark, ed. (London and Nashville, 1958) 2: p. 741; J. Thomas Scharf, *History of Western Maryland* (Philadelphia, 1882) 2: p. 1466; and James W. Thomas and T. J. C. Williams, *History of Allegany County* (n.p., 1923) 1: p. 546, and 2: p. 1094.

[23] Robert Graham, a stockholder and later director of the company, was about to take over management of the accounts at Lonaconing, including store affairs, the ordering of supplies, and payment of the employees and contractors.

[24] Bowers built the forge in 1830 and the furnace in 1835. Both stopped running in 1856. James M. Swank, *Introduction to a History of Ironmaking and Coal Mining in Pennsylvania* (Philadelphia, 1878), p. 32.

June 22.—In the morning P.T.T. and J.H.A., accompanied by Mr. Schmidt, revisit the new limestone bed, where also they make observations with the barometer and the thermo-barometer of Mr. Wollaston.[25] The results seem to demonstrate the feasibility of a road on the route mentioned of yesterday. In No. 1 drift a confusion among the workmen, and two discharged who demanded $2 for 8 hours' work. No. 3 timbering finished and well done. An accident happens to one of the miners, who fell on the priming needle.[26]

June 23.—The miners from No. 1 discharged. To avoid confusion, drift No. 1 is hereafter called drift *A*; No. 2, drift *B*; and No. 3, drift *C*.

June 25.—J.H.A. confined by an accident. Houses floored in all the day, and Mr. Pauer directed to inquire in Cumberland for a plasterer. The trials with the patent fuse very satisfactory. Mr. Wilson enclosed Mr. Bruinell's list of the miners from Wales.[27]

June 26.—Sawmill working at water wheel stuff. It averaged during the month of May 550 [feet] per day. [Letter from] King, Swope and Co. [of Bedford County, Pa.] offering bar iron at 5½ cents per pound, cost of hauling not known.

June 27.—Two plasterers arrive, one from Frostburg, the other from Cumberland. A contract made with the former to finish the plastering out and out, furnishing everything except lime and sand and one tender, for $36. Work to be done in one week. Cars badly wanted and carpenter proceeding with them. Drift *A* commenced on the night of the 26th by the old hands without altercation as to the old prices. Engineer taking engine frame apart ready for erection on the foundation. Powder magazine going on. Determined to sheathe its arched roof with a layer of common bricks and afterwards to *pitch* it over. (Ha-ha!!!) John and William Hopkins, brother and son of David [the founder], arrive this evening, as well as a new blacksmith, Luther Shendin, from Baltimore.

June 28.—Powder magazine bids fair to verify the unfortunate *jeu de mot* in the entry of last night by cracking the counterwall and threatening to upset altogether. An investigation of the conditions of equilibrium of the arch and walls of the magazine at the point of springing of the arch according to the formula of Gregory[28] shows that the piers are strong enough. The cracking is to be attributed to the heavy shower of rain which fell last night.

June 29.—The limestone at Ayres' going on well, turning out upwards of 6 feet thick. Mr. Pauer directed to run a line of levels so as to determine a point about 30 feet above the base of the 14-foot coal, which is presumed to be the level of the limestone bed on Dug Hill. [Alexander hoped to find an extension of Ayres' limestone on the company's property.]

June 30.—Mr. Alexander's wound still painful, but he goes to Frostburg in the afternoon with Mr. Graham. P.T.T. follows.

JULY 1–DECEMBER 31, 1838

During the second half of 1838 the blast furnace and its auxiliary structures were completed, the steam engine and blowing machinery were set up, and all the blast pipes were laid. On the east side of the big stack were the engine house and the boilers with their own furnace and chimney. At the two back corners were the hot air furnaces.

The machinery purchased from the West Point Foundry was only a part of the blast system. Its function was to force cold air through cast-iron pipes to the two hot air furnaces where it would be heated to 700°F. in a nest of smaller pipes before passing through the tuyeres into the blast furnace. Coal burned in chambers beneath the pipes in the furnaces provided the heat. The hot blast, first introduced in England in 1828, made possible a saving of thirty to sixty per cent in fuel and speeded up the smelting process.[1]

Most of the blast pipes and the iron fixtures for the hot air furnaces were ordered from three foundries in Baltimore: Messrs. J. Barker & Son, Ellicott's Iron Works, and Ross Winans.[2] These manufacturers delivered their castings to William Alexander, the company's agent, for forwarding. In November he had sent on three wagons with pipes from Barker's works.[3] Additional orders from Barker in March 1838 amounted to more than 1,800 pounds.[4]

While the masonry was being finished and the pipes installed, contractors worked at digging out and grading the top house yard, coke yard, and ore roasting yard, so that all would be ready for making iron early in 1839. Since work was slack on the

---

[25] William Hyde Wollaston (1766–1828), English chemist and natural philosopher, *Dict. Natl. Biog.* See text following entry for May 27 concerning use of the barometer for determining altitudes.

[26] The priming needle was a thin (¼″) tapering copper rod used for making a hole to contain the black powder or fuse used in setting off a blasting charge.

[27] In view of the eccentricities of spelling in the journal, the reference may be to the English engineer, Isambard Kingdom Brunel. Certainly in 1838 Brunel was in a position to do Alexander this favor. The Great Western Railway was under construction at the Bristol end, and Brunel was also building the Merthyr and Cardiff (Taff Vale) Railway through the heart of the South Wales coal and iron country. Furthermore, he was using miners for his tunnel work. One might reasonably assume some correspondence among civil engineers in the 1830's when their profession was in its infancy.

[28] David Gregory (1661–1708), Scottish mathematician and astronomer, whose works contained "demonstrations of various properties of the catenary curve, with the suggestion that its inversion gave the true form of the arch." *Dict. Natl. Biog.*

---

[1] Alexander, *Manufacture of Iron*, p. 92; J. E. Johnson, Jr., *Blast-Furnace Construction in America* (New York, 1917), pp. 190–191; Tomlinson, 2: p. 83; Overman, *Metallurgy*, pp. 419–420, 528–529; and Overman, *Manufacture*, pp. 441–442.

[2] Journal, Oct. 30, 1837, July 21 and Aug. 31, 1838. Messrs. J. Barker & Son owned the Curtis Creek furnace and two foundries near Baltimore. Their foundries used about two-thirds of their furnace product for making castings. Alexander, *Manufacture of Iron*, p. 88. Ellicott's iron works at Baltimore had the reputation of making the best cast iron pipes "in this or any other country." Hunt's *Merchants' Magazine* 5 (1841): p. 54. Winans, as we have already seen, also supplied tram wheels for the company.

[3] Journal, Nov. 13 and 15, 1837.

[4] Daybook, Mar. 16 and 23, 1838, Welch and Alexander Record Books, Md. Hist. Soc.

canal, the company had no trouble in finding bidders for the various other jobs it let to excavating contractors during this period: foundations for the sawmill and store, the sawmill dam and millrace, and the lumber road to Beatty's Plains.

The company expanded its corps of surveying engineers to locate routes for the lumber road, the turnpike to the National Road, and a railroad to the proposed shipping depot on the Potomac at Buck Lodge.

The increase in disciplinary problems as the year progressed led to the adoption of a strict set of rules for the government of the Lonaconing population and to the institution of nightly patrols by bailiffs. The miners recruited from Wales proved particularly troublesome, causing Alexander to remark, ". . . upon their accession we have no great reason to congratulate ourselves."[5]

Tyson keeping journal:

July 3.—Top house arch completed.

July 4.—The masons, carpenters, and part of the laborers at work. A row between the natives and some of our Welshmen at Frostburg.

July 5.—Plasterer arrives from Frostburg and cuts timbers for laths. About half of the Welshmen not returned to their work. [Stockholder Roswell L. Colt and his son visit Lonaconing.]

July 7.—Duff Green, Major Douglass, and the two Howells make their appearance. Mr. Yonger [commences] setting the frame of the engine. Owing to the excessive heat of the weather, the miners above ground suffer considerably. A number have diarrhoea.

July 9.—Two miners commence a windway in the coal between ores 2 and 3 to furnish air to C drift.

July 10.—The weather continues excessively hot. Thermometer ranges from 85 to 90 in the daytime in most places. On the patch [open work] it is about 130°. Several more men sick in consequence. Recommend that they commence work an hour sooner and work one hour later, and take two hours' rest more during the heat of the day. The plasterer having trifled away a week and done nothing, [I] ordered two of the masons to finish the plastering of the stone houses.

July 11.—Rupert sent to Frostburg with the carryall for laths. The grey mare runs off with the carryall and breaks it to pieces. He, however, borrows a cart and brings part of his load down. The steam cylinder is put on the frame.

July 12.—A load of flour and bacon arrives from Virginia, and a load of plank from Westernport. Mr. Alexander returns from Frostburg. [Letter to] A. Duvall ordering the lath and shingle machines [for the new sawmill which Duvall will build.]

July 13.—Our team brings a load of laths from Frostburg, as well as the wreck of the carryall.

Alexander resumes journal:

July 16.—Millwrights, five in number, arrive from Baltimore. Railroad lumber entirely out from William Shaw's failing to send as contracted for.

July 17.—Digging for the store foundation commenced today, and a bed of good sand developed. Millwrights who arrived yesterday evening commence getting out

timber today. One at work on a shed to frame under. Letter to Matthew St. Clair Clark about the right of way for the turnpike.[6]

July 19.—John Becker, miner, discharged for drunkenness. Powder used for the first time from the magazine, to which it had been removed on the 18th.

July 20.—Determination of elbow pipes for the blast and hot air furnaces. [Here are diagrams and further discussion of arrangements of pipes and grates.]

July 21.—Professor [Benjamin] Silliman and son come here this morning and spend the day.

July 23.—Foundation of boiler furnace commenced. Contract made for hauling stone from the ford for the store, boiler furnace, etc., at 50 cents per perch laid for 500 perches or thereabouts.

July 26.—Mr. T. S. Alexander arrives. Letter to Ross Winans ordering 200 pounds of cast iron borings for cement[7]; two fire doors and jambs for hot air furnaces, between 20″ and 30″ square or according to the patterns; 2 small doors of 15″ square for manholes to the hot air pipes; a mandrel for welding or riveting the blast pipes [illustration given]; and the valve fixtures for the blast pipe at the egress of the hot air furnace. 2 in number [illustration given, followed by a description of the way the valve fixtures will work].

July 27.—Weather is extremely hot. Hearth being cleaned out and prepared for receiving the hearthstones. A new contract made with William Ledley in regard to the brickmaking: viz. at $4.50 per thousand, he finding everything and delivering the bricks anywhere on the flat and for the stack chimney. None but good bricks (perfect) to come under the contract. Also the wood already cut to belong to him for this purpose.

July 29, Sunday.—C drift railroad takes fire, but is extinguished without difficulty.

July 30.—Duvall [sawmill contractor] arrives today, and Mr. Richard Serpell to whom employment [as surveying engineer] is given to commence from 31st at $2.50 per diem. Steam engine foundation finished, and diggings commenced for chimney stack. Setting the hearthstones commenced today.

July 31.—Mr. Serpell set to work at leveling the proposed lumber road over by Jacob Koontz'. J.H.A. occupied in giving the corners of the store building, and in the afternoon with P.T.T. in working out the proposed position of the sawmill, which, however, does not suit Mr. Duvall's notions. Hearth progressing until at 6½ the gib [of the crane] at the top capsized without, however, doing any material damage. P.T.T.'s family arrive today, and he and R.G. are semidomesticated in the new house.

August 1.—J.H.A.'s family came. Foundation [of sawmill] commenced today.

August 4.—Four hands go away, leaving a great deficiency in laborers.

August 6.—A new yoke of oxen purchased.

August 7.—Steam engineer finds that the valves in the steam cylinder have been wrongly fitted, and he has therefore to take the concern apart.

[6] Clark was president of the Maryland Mining Company, whose lands were a little north of Commonwealth. A turnpike to the National Road would cross the Maryland company's property.

[7] The iron borings, scrap from manufacturing operations, were an ingredient of the cement used to make the joints of the blast pipes airtight. One formula for the cement was: 5 parts iron borings, 1 part fine clay, moistened by vinegar. An alternative formula was: 60 parts borings, 1 part sal ammonia, 6 parts clay, the whole moistened by water. Overman, *Metallurgy*, p. 420.

[5] GCC&I Co. Report 1839, p. 6.

August 10.—August [Weisskettle] and [John] Schwenner dressing tymp stone. Engineer engaged in fitting a new awkwardness between the feed pipes and eccentric shaft of engine.

August 11.—Letter to R. Winans enclosing a drawing of coke barrow wheels and ordering 6 pair, also inquiring after a patternmaker mentioned by Duvall as having worked with Winans. Mr. Charles Preuss, engineer, arrives.

Charles Preuss, trained as a surveyor and cartographer, was born in Germany in 1803. He migrated to the United States in 1834 and almost immediately found work under Ferdinand Hassler in the United States Coast Survey. No doubt Preuss and Alexander met in the course of Alexander's work as topographical engineer for the state of Maryland. A moody man, Preuss was unhappy at Lonaconing, where he felt that his talents were not fully appreciated.[8] After leaving the George's Creek company in 1839, Preuss was engaged by John C. Frémont as a cartographer on the 1842, 1843, and 1848 expeditions. When the last Frémont expedition failed, Preuss remained for a while as a surveyor in California, became ill, and returned to his home in Washington, D.C., in 1850. Increasingly melancholy and unable to work, he hanged himself in September 1854.[9]

August 13.—Watkins Pritchard, blacksmith, discharged.

August 14.—The tymp stone fitted this afternoon. Carpenters engaged in completing the shed erected for store goods by the old office, which is now given to Mr. Preuss as a residence, and in fitting up his house.

August 15.—Hearthstone finally fixed and rammed down. Hands engaged in dressing and fitting the upper course of the boshes. Location given for stable. Gauging made of the stream which flows from the coal mine to ascertain its sufficiency for the steam engine. [It is found to supply 150 gallons an hour, $\frac{2}{3}$ of the amount needed.]

August 16.—Three masons not at work from sickness of themselves or families. The subject of the dam and race taken up, and resolution taken to prosecute it vigorously. At night letter from W. Alexander.

William's letter informed J.H.A. that he was forwarding by railroad a dray load of machinery just arrived from New York. He also announced that he had on the day before sent down a load of some of the iron ordered from Winans, as well as some powder. In order to secure lower freight rates, William had begun early in July to ship the company's goods via Winchester, Virginia. Routing was on the Baltimore and Ohio to Harpers Ferry, thence on the Winchester and Potomac Railroad to Winchester, where Isaac Paul, a commission merchant and forwarding agent, picked up the shipments and transported them by wagon to Lonaconing.[10] Wil-

liam's letter had a significant postscript: "Reports unfavorable to your undertaking as to the quantity of ore are pretty current. Evan T. Ellicott told me he was told you had enough to keep one furnace at work, but not more."[11]

August 17.—Work as usual in the mines, except that a disturbance arises in consequence of an allegation made by the Welsh blacksmith who takes the place of Pritchard mentioned in the journal of the 13th that certain of the miners threatened his personal safety if he continued to work. The only hand who was directly accused was ordered to be discharged, but upon further examination the allegation was not considered proven, and due submissions and protestations having been made, the superintendent agrees to overlook this case. Clearing made for the dam foundation, and an approximate draft made of its extent and dimensions. Letter to Joseph Dilly offering him a contract for the dam and part of the turnpike.[12]

August 18.—The ground for the dam is laid out according to the following proportions as shown in the sketch below. [The sketch shows a dam 72 feet wide, and a millrace 15 feet wide.] Lightning rod in powder magazine erected this evening.

August 19, Sunday.—At night Mr. Walters from Farrandsville presents himself.[13]

August 20.—Mr. Walters perambulates the works in company with P.T.T. and J.H.A. Dilly arrives, and the job being shown him, an agreement is made for the excavation of the dam foundation at $1 per cubic yard. In the afternoon Mr. Pauer commences the railroad location levels.

August 21.—In the morning the consideration of the interference of the hot air well with the position of the regulator [for blast machinery] is taken up, and after deliberation it is resolved to leave our own machinery as we had fixed it, and then to alter the steam engine fixtures to suit. William Hopkins removes the drying furnace [for the masonry] in the east pier.

August 22.—Bricklayers progressing with the boiler furnace. Northeast wing wall finished for the present at night, and railroad laid so as to bring the stones for the inwalls from the patch. In the afternoon the blast pipes are placed in their position in the back alley, prior to fixing the center of the regulator. Mr. Preuss runs the line of the millrace, and Mr. Serpell levels the foundation of the dam and also the line of the outer bank of the race.

[8] Preuss to Hassler, Dec. 17, 1838, Hassler Papers, Manuscript Division, New York Public Library.

[9] Charles Preuss, Exploring with Frémont, trans. and ed. by Ervin G. and Elisabeth K. Gudde (Norman, Okla., 1958), pp. xix–xxix.

[10] W. Alexander to Isaac Paul, July 6 and 17, 1838; and W. A. to R. Graham, July 7 and 26, 1838. Letter Book 1,

Welch and Alexander Record Books. For lists of the items shipped, see W. A. to Paul, Aug. 13, 14, and 15, 1838.

[11] W.A. to J.H.A., Aug. 14, 1838, Alexander Papers.

[12] Dilly, the owner of a large amount of land in Allegany County, was one of the contractors for maintenance of the National Road. Thomas and Williams, 1: p. 569; and Scharf, 2: p. 1328.

[13] William T. Walters was manager of an iron works at Farrandsville in Clinton County, Pa., on the west branch of the Susquehanna River. The furnace, built to use coke, was blown in in 1837 and probably ran until 1839. During this period it produced about 3,500 tons of iron. Since the ore had to be brought from a considerable distance (as much as 100 miles), the cost of manufacture was high. After sinking about a half million dollars in the enterprise, the owners decided to dispose of the works. James M. Swank, History of the Manufacture of Iron in All Ages (2d ed., Philadelphia, 1892), pp. 201, 369; and Walter R. Johnson, Notes on the Use of Anthracite in the Manufacture of Iron (Boston, 1841), pp. 6–7.

August 23.—Masons finishing off northeast wing wall and at work on the foundation of the top house. One of the hands engaged by the contractor for the sawmill foundation is considerably injured by the falling in of the bank upon him.

August 24.—Hauling of stone from patch still continued, but in consequence of thoughtlessness of Mr. Schmidt a trouble is like to grow from it. The circumstances were that on the 23rd, when the work on the patch was done by the Welsh miners under a temperature of 120° or thereabouts, their request through Mr. Hopkins to be allowed the use of a small quantity of wine whilst at their work was granted, and 1½ gallons accordingly placed with Mr. H. for disposal through the day and part of the night that they worked. When today came, the Welshmen were not in time, and therefore German hands were put at the same work. They preferred the same petition, and with success, but instead of wine being given to the patch hands only, the article was also distributed among other Germans unloading near the stack beside the masons, to whom no liquor was offered. The latter consider this a great grievance, and when the bell rings, do not go to work. They are told by P.T.T. through Mr. Harris that they shall either go to work immediately or go home. They prefer the former, and matters go on smooth again. [Blast] pipes are partly fitted this evening.

August 27.—The contractor (Renninger) arrives with some of his hands to level [grade] the furnace yard. Mr. [Christian] Renninger's contract is at 27 cents the cubic yard of excavation. Mr. Dilly and [William] Coombes, contractors for the mill dam excavation, arrive but do not commence work. Mr. Hopkins has agreed with Buskirk for the tenancy of the house on lot 3720, late Mr. Steele's residence, at $50 per annum during a term of 5 years, the lease to be broken at any time after notice of one month on the part of Mr. Hopkins. Mr. Preuss, surveyor, and Pauer, leveler, commence the location of the turnpike. Regulator [for the blast machinery] placed on the flat by the place for it, according to the plan formerly determined on, and brick foundation laid. Masons (two) commence with the inwalls [of the blast furnace].

The interior of a blast furnace is often described as two truncated cones attached base to base. Above the hearth the lower cone widens upwards in a gentle slope (the boshes), converting the interior shape from the square of the hearth to a perfect circle at the widest part of the shaft (the bosh) where the two cones join. The boshes are laid in courses of solid stones closely joined to the rough wall of the stack. The company's stonecutters had begun working to quarry out and dress the bosh stones early in March 1838. Above the bosh the upper cone rises to the narrow mouth of the furnace, which in the journal is called the trunnel head. The upper cone is the inwall, or lining, of the furnace. It is not attached to the rough walls, which have, however, been shaped to contain it. Between the rough walls and the inwall is a space six to eight inches wide loosely filled with broken stones or bricks. The inwall itself rests on the piers and arches previously described. Ideally the inwall is made of firebrick. It is evident from the journal entry of September 17, 1837, that Alexander in-

tended to use firebrick for the inwall and for lining the rough walls above the bosh, but when he found that he would not be able to secure firebricks, he settled on a lining of sandstone, which in his own words answered the purpose "tolerably well." If sandstone is used for the inwall, it must be of a fine-grained white variety which will not crack and which will stand the heat of the fire. The stones must be cut and dressed according to bevel and circle, and must be laid in courses of equal thickness. The top five courses should be of firebrick or of well-burnt common brick.[14]

The stonecutters Schwenner and Weisskettle worked at dressing inwall stones from June 20 to August 20, 1838. The stones were hauled down to the stack on August 23 and 24, and on August 27 the masons began the inwalls. Work on this part of the furnace was finished on October 5.

August 28.—The greater part of the new [steam engine] machinery was received yesterday but not used today. The regulator as far as riveted by Farron is placed upon its seat. Coombes and Dilly commence on the dam foundation today, and word is brought that Mr. Buskirk has prohibited the timber hewers from cutting on the other side of the creek, which prohibition is, however, disregarded.

August 29.—Mr. Serpell leveling for the furnace yard. Ultimate limits not yet determined. The connecting pipe of Kemble for the new arrangement of the boilers is found entirely wrong, and measurements taken for a new one.

August 31.—Reverend Mr. Harris arrives after dinner, and at night celebrates Divine Service.[15]

September 1.—Bricklayers going on with the boiler furnace and getting it ready for the boilers. The miners still hold out about prices [contract] which it is proposed to lower.

September 3.—Miners still unsettled, but told to make up their minds by Tuesday [next day]. Boilers put in their place, to the injury of Yonger's foot.

September 4.—Sixteen [miners] are paid off in the afternoon who cannot take up the work at the prices offered. Yonger is unable to work today, but the riveting of the regulator is carried on by John Jones. The bricklayers are sent to work on the wing wall and the south foundation of top house, which is started today. Consideration of zinc roof taken up, and preparatory sketches made from measurements taken in the evening.

September 5.—The refractory miners are finally paid off. Consideration of roof continued, and letters written to Mr. Lorman [Wm. Lorman & Sons, Baltimore merchants] to contract for 10,000 square feet or 9,000 pounds of zinc, and to Mr. Wilson to inquire for a zinc plumber who can come up and do it.

14 Alexander, *Manufacture of Iron*, pp. 117–118; Tomlinson, 2: p. 77; Overman, *Metallurgy*, p. 507; Benjamin Silliman, *Extracts from a Report Made to the New York and Maryland Coal & Iron Company* [sic] (London, 1838), p. 44; Overman, *Manufacture*, p. 156; and GCC&I Co. Report 1839, p. 7.
15 The Reverend Mathias Harris was rector of Emmanuel Parish of the Protestant Episcopal church in Cumberland from 1837 to 1841. *Journal of a Convention of the Protestant Episcopal Church of Maryland, 1837* (Baltimore, 1837), p. 13; and Scharf, 2: p. 1413.

A metal-clad roof was a necessity for buildings so close to the furnace. Sheet zinc cost about the same as tin, and was cheaper than iron or copper. All of these metals were being used for roofs in the 1830's.[16] Zinc was available from both domestic and foreign manufacturers.

The company's zinc was shipped from Baltimore, rolled and packed in casks, each containing about 1,100 pounds. William Alexander admonished the forwarding agent at Winchester that if extra coopering became necessary, the utmost care should be taken "to prevent any nail being driven in so as to enter or injure the zinc."[17]

September 6.—Dam foundation going on preparatory to the masons commencing work. The boiler stack scaffolding finished all ready for use.
September 7.—John Hopkins [who replaced Steele as underground agent] reports only 615 tons [of ore] on the place.
September 8.—Dam going on till 12 o'clock, when Coombes' hands leave for camp meeting, having worked the previous night at 12 o'clock. [Thomas] Layson [miner] agrees to go to work on Monday at the reduced prices, which offends the other recusants very much.
September 9, Sunday.—This day is only commemorable because of an outrage having been committed by some person or persons unknown on the house [of] Thomas Layson, the miner about whom the entry was made yesterday. A reward of $100 was offered by the superintendent for any information leading to the detection of the offender. At night two additional watchmen stationed privately in Layson's house. No further disturbance.
September 10.—Layson, M. Rowland, and D. Williams, recusants, commence work today at the reduced price, and Reverend Jones, supposed ringleader, applies to be taken back on promise of good behavior, which Mr. Hopkins agrees to. [Among the Welsh miners, D. Jones, Levi Harris, and Richard Harris were lay preachers.] Pattern drawn and sent to F. H. Oliphant for a support above the fill-in doors of the chimney stack. [A sketch follows.]
September 12.—J.H.A.'s family leave Lonaconing for Baltimore.[18] Dam going on by the masons, who are taken off at dinner time, having finished. John Hopkins gives in some details as to the quantity of work which can be done in the No. 4 ore, shewing that at two [dollars]

per diem the ore can be got out at $2.50 per ton. Foundation of the south hot air furnace laid out and commenced with two hands.
September 13.—In the morning Wagner [contractor] is shewn the race bank, which is offered to him at 25 cents per cubic yard. He asks, however, $37\frac{1}{2}$, which is considered too much. Coombes appears and agrees to take the contract of the top house yard at 27 cents per cubic yard.
September 15.—Four of the hands sent by the *Tiberias* arrive.

The 109-foot barque *Tiberias* arrived at Baltimore on September 11, 1838, at the end of a voyage of forty-six days from Newport, Wales, bringing seventy-five passengers "destined for the George's Creek Company." She also carried a cargo of 3,689 pieces of railroad iron. All but six of the passengers were members of the eleven families on board—husbands and wives with sons and daughters ranging in age from infancy to the early twenties. On the passenger list, twenty-eight of the males were described as "colliers." Of these, nine were boys ten to sixteen years old. Two adult males, listed as "founders," were employed as keepers. It is interesting that Sarah Davies, aged thirteen, was also classified as a "collier," but there is no indication that she, or any other woman, ever worked in the Lonaconing mines.[19]

September 16, Sunday.—Divine Service in afternoon by Mr. Harris.
September 17.—Bricklayers leave the boiler stack at midday to commence the north hot air furnace, having also reached the point beyond which no mortar or bricks can be carried by any of our hands. Yonger going on with his fixings [steam engine] but slowly.
September 18.—Mines as usual. Some little difficulties still about prices, etc., which, however, appear to be adjusted. One of the new *Tiberias* men arrives today, having come in the stage. Wagner, contractor, commences today at the race, grubbing and clearing. Clarke, the chief blacksmith, is discovered today to be drunk, and upon examination tomorrow will probably be discharged. Mr. Pauer copies today and yesterday the new rules of the superintendency. Contract made for the delivery of building stone for the molding house, store, and sawmill at 75 cents per perch measured on the wall. Store wants 80 perches, molding house 300, sawmill 600 perches.
September 19.—Three more hands by the *Tiberias* arrive this evening. The fight which was represented as having occurred yesterday evening results in the discharge of the two hands engaged in it. The north [hot air] furnace is sadly interfered with by the blast pipes, which renders necessary some little alterations in the position of the door. Rupert, who goes to Frostburg for flour, returns without being able to procure any. Determined to write to Mr. Hopewell of Moorefields [John Hopewell, who managed J.H.A.'s land in Moorefield, Virginia] for wheat and flour, etc. Letter from W.A. at night announcing the employment of hod carriers, two in number, and mentioning about a bricklayer who wants $1.37\frac{1}{2}$ per thousand without tender.

[16] For conflicting opinions regarding the durability of zinc and the possibility of installing it with leak-proof joints, see L. D. Gale (New York Univ.), "On the Uses of Zinc for Roofing of Buildings," *Mechanics' Magazine*, Mar. 1836: pp. 163–166; Gale, "On Zinc Roofing," *Amer. Jour. Science* Series 1, 32 (1837): pp. 315–319; and A. Caswell (Brown Univ.), "On Zinc as a Covering for Buildings," *ibid*. 31 (1837): pp. 248–252.
[17] W. A. to Isaac Paul, Sept. 22, 24, and 27, 1838, Letter Book 1, Welch and Alexander Record Books. See also account book, *ibid*., entries Sept. 22 and 25, 1838.
[18] Mrs. Alexander had been ill since the end of August. William Alexander sent down medicine on August 31, together with directions for its use. On September 4 he wrote that he and a female relative of Mrs. A. were ready to start for Lonaconing. On September 7 they were at Hagerstown with a driver, carriage, and four horses, and sent word that they expected to arrive at Lonaconing as early as possible on Monday morning, September 10. W. Alexander to J. H. Alexander, Sept. 4 and 7, 1838, Alexander Papers.

[19] *Tiberias* passenger list, giving name, age, and occupation of each passenger, National Archives; and Baltimore *American*, Sept. 11 and 15, 1838. See Appendix A.

September 20.—Bricklayers (2 new ones) at the hot air furnaces. An investigation of the weight of cast iron borings necessary for the cementing of the blast pipes is made at night. [The calculations and a sketch follow, indicating that about 663 pounds of cement will be needed for the 96 joints of the blast pipes. See footnote 7 for composition of the cement.] Inwalls consume all the stone ready for them. Hopkins' house not laid out because of the entire occupancy of the ground by the millwrights. Letters to Mr. Hopewell, offering $7.50 per barrel for flour and $1.50 for wheat delivered here, and requesting his agency in the purchases; [and] to W. Alexander, offering the bricklayer $1.37½ per thousand without tender or labor.

And thus endeth the chapter of proceedings for thirteen months of calculation, contrivance, and makeshift, with a small sprinkling of success and good hope for the time to come—all of which will in due time come to light in the next and succeeding volumes of the Lonaconing Transactions. [End of Volume 1]

Volume 2—Lonaconing Transactions

RULES OF THE LONACONING RESIDENCY*

The superintendent of the works of the George's Creek Coal and Iron Company at Lonaconing has prescribed the following rules for the government of all persons in the service of the company.

1. Every department of the works, whether mining, carpentry, blacksmithing, masonry, or digging of any kind is under the charge of a particular manager, by whom all the hands in the respective departments are immediately superintended and to whom every hand in his particular department will be required to pay entire respect and obedience. These managers are selected in the discretion of the superintendent, and report to him the state and progress of the work in their particular charge. They also report the time and wages of the hands in their employ, by which report the payments for wages are all governed.

2. Every person in the employment of the company will be required to be present at work on every day in the year excepting Sundays and Christmas Day, and the hours of employment (excepting in special cases which the superintendent allows in his discretion) shall be from sunrise to sunset, with such intermission for meals as shall from time to time be appointed.

3. Signals are given, by tolling the great bell of the company, of the hours for beginning and leaving work, and every person who is not in his place in the proper time will be expected to account for his absence to his immediate manager.

4. Absences from work will be punished by abatement from the wages at the discretion of the manager; and where such absences are frequent, or without sufficient excuse, the individual who is guilty will be discharged and reported by the manager to the superintendent.

5. No distilled spirituous liquor shall be sold on the grounds of the company, nor shall any distilled liquor be used by any person whilst he is actually engaged at work for the company. Intoxication at all times and places is strictly prohibited. The managers of the several departments are required to enforce the strict observance of this rule by dismissing immediately from service any person or persons under their respective charge who may be guilty of a violation thereof. The superintendent

requires its observation also on the part of the managers themselves, and all others in the service of the company are likewise in such case directly responsible to him.

6. All brawling, quarreling, fighting, and gaming are prohibited. The firing of guns, which has been more frequent of late than usual in the valley of the works, is also, as dangerous and unnecessary amusement, forbidden in future. The managers must aid in the enforcement of this rule by reporting to the superintendent all violations thereof which may come under their knowledge.

7. The company expects all persons in its service to observe an orderly and decorous conduct towards each other; and if any controversy should arise between such persons, the superintendent will, on application, use his good offices and authority to reconcile the parties on just and proper grounds.

8. Every person in the service of the company, whether employed by the day, month, or year, shall be liable to be dismissed at the discretion of the superintendent, and on his dismissal shall leave the grounds of the company.

9. In like manner the superintendent claims the power of determining at any moment any lease or renting of the company's grounds or houses, whether the same be made by the day, month, or year; and it is expressly declared that no tenant of the company shall board or lodge any person who may be discharged from the service of the company, nor shall he be permitted to engage in any manner in buying or selling distilled spirituous liquors on the company's grounds.

When work is contracted to be done by the job, it is understood that it is to be subject to the acceptance or rejection of the contractor; and all persons entering into such contracts are to be subject to the operation of these rules, and to be liable to have their contracts rescinded for a violation thereof.

10. Monthly settlements will be made with every individual in the regular employment of the company. From the sum due him for labor will be deducted the amount of the accounts against him at the store, mills, and post office, and his contributions for the doctor and school fund as hereafter mentioned, and the balance will be paid him in money or a check on Baltimore or Cumberland at the discretion of the superintendent.

11. For promotion of the general health a physician is settled at Lonaconing. Every able-bodied man in the service of the company, who will contribute out of his wages a monthly sum of fifty cents for the support of the physician, will be entitled to medical advice and assistance for himself and family without further charge.

12. The physician will make a daily report to the superintendent of the general health of the village, and the numbers of deaths and births within its limits; and he will also from time to time specially report all nuisances which may affect the health or comfort of any portion of the inhabitants.

13. Every tenant of the company is expected to preserve neatness and cleanliness about his premises, and more especially to pay attention to the condition of the sty or other places where he may keep his hogs and pigs. The superintendent will insist on the observance of this rule and will prohibit the keeping of hogs or pigs by any person who may disregard the same.

14. After the first day of January next [1839] no dog shall be kept on the company grounds without special permission of the superintendent, who may withdraw the same at any time in his discretion.

September 21.—The rules on the preceding pages having been drawn up with care by Mr. T. S. Alexander (a few slight modifications only being made) were this

---

* Reprinted from Katherine A. Harvey, *The Best-Dressed Miners* (copyright © 1969 by Cornell University). Used by permission of Cornell University Press.

day adopted, and the carpenter having made a straining frame [bulletin board], they have been therein inserted and hung up for the benefit of the public.[20] New hands by the *Tiberias* arrive. Schmidt's hands digging for the roasting yard on the northeast of inclined plane, and also hauling stone for the east pier of dam, in the discharge of which they are interrupted by Buskirk. Upon investigation it is found that they are actually on his grounds, and they are therefore directed to quit that and get the stone from Hills Run, undoubtedly on our own ground. It is reported at night that Buskirk has interfered there also.

September 24.—Two *Tiberias* men with families, and two Pottsville men ditto, make their appearance. Steam engine machinery now all arrived, and bricklayers, having finished the north hot air furnace, are at work on the boilers.

September 25.—Four hands [miners] leave the works on account of low prices, but the places of two are supplied by old hands who went away at the last commotion. Mr. Hopkins, who had most imprudently been contravening the rules by selling liquor, is reproved by Mr. Tyson and promises amendment.

September 26.—Four families by the *Tiberias* arrive. Inwalls progressing slowly in consequence of the rain.

September 27.—Mining as usual, except that there appeared difficulty likely to occur from the newcomers not being satisfied with $1.50 per diem. After a representation to them, however, of the circumstances of their contract, etc., by P.T.T., they are satisfied and go to work. Letters enclosing an advertisement for the letting of the turnpike road on the 29th October written to the *National Intelligencer*, Baltimore *American* and *Patriot*, *Frederick Herald*, and Cumberland *Civilian*.[21] Also to Charles B. Shaw, proposing to him to come to Lonaconing as the superintendent.[22]

---

[20] Presumably the rules were displayed in three languages. J.H.A. himself translated them into German and Welsh. J.H.A. statement of 1850, Alexander Papers. It is interesting to note that the Lonaconing rules were concerned with the behavior of the workers and the preservation of respect for authority at all times, not merely within working hours. Compare the rules of a contemporary English establishment governing employees only while they were at work. Arthur Raistrick. *Dynasty of Iron Founders: The Darbys and Coalbrookdale* (New York, 1970 [Reprint]), pp. 298–299.

[21] The advertisement invited proposals for "the graduation, bridging, and stone covering of a macadamized turnpike road about 9 miles in length from the company's works to the National Road." Baltimore *American*, October 1, 1838.

[22] Charles B. Shaw (1800–1870), a civil engineer, was employed by the Virginia Board of Public Works from 1831 to 1834 as assistant to Colonel Claudius Crozet in locating and building the Northwest Turnpike connecting Winchester and Parkersburg. In 1835 Shaw succeeded Crozet as principal engineer for the Board. In July 1836 Shaw informed the board that he intended to resign in March 1837. None of the available records gives any reason for this decision; so one surmises that Alexander may already have approached Shaw on the subject of the Lonaconing superintendency. The two men met at least as early as 1834, when Alexander, as state topographical engineer, was mapping the Eastern Shore of Maryland, and Shaw had been appointed, at the specific request of the governor of Virginia, as a member of a commission to study the Chesapeake and Delaware canal. The Alexander Papers contain a receipt, dated June 23, 1834, for payment on behalf of "The Commission for the Survey of the Coast of Virginia, Maryland and Delaware" for board and lodging of C. B. Shaw, J. H. Alexander, and two other persons. It is also possible that they met again in 1836, while Shaw was gathering data for his

September 28.—Visit from Mr. Oliphant. Mr. Duvall arrives, and Strebig [miller], who has made purchases of 54 barrels of flour—40 at $8 delivered, and 14 at $8 and the carriage.

September 30, Sunday.—The peace of the night previous having been broken by a riotous assemblage at and before the house of William Koenig, resulting in the demolition of the windows of said house by some person or persons unknown, the entire day was given to the examination by Mr. William Shaw [magistrate] of sundry witnesses to the transactions, who, however, could give no satisfactory information as to the persons of the individuals active in the outrage. The result of the evidence was: that William Koenig was guilty of selling spirituous liquor on that night to several persons, though he swore he did not; that William Zimmerman was guilty of fighting, but whether as attacker or attacked could not be made out satisfactorily; that William Farley was guilty of fighting; that Owen Richards was guilty of being manifestly drunk; and that David Rees (discharged) was probably the active agent in this Fronde [uprising]. It also appeared that William Mitchell and John Powell knew more about it than they chose to disclose. At night by way of flourish 8 guns were brought to the house and stacked up loaded, and a secret patrol perambulated the streets and byways to assure us of quietness and also if possible to meet with the rioters of the preceding night, but without success. Everything continued orderly and quiet.

October 1.—Owen Richards and William Farley are discharged for their share in Saturday night's transaction. John Powell is reprimanded and permitted to remain only on account of his wife, and William Mitchell proves by fresh testimony that he did not aid in throwing any stones. He is therefore allowed to remain. Koenig, brakeman, is discharged also, and Zimmerman. At night George Blatter and William Howell are appointed bailiffs and directed to watch together for three nights and then to halve the night between them. Letter to the adjutant general [of Maryland], asking the loan of 25 muskets.

The letter to the adjutant general reads in part: "We have here now so many people of different nations that in spite of all our regulations and efforts we cannot always keep the perfect peace among them which is necessary to be preserved, and it occurs to me that the presence of a few stand of arms would aid materially in causing the law to be respected. Particularly now, when the contractors engaged on the turnpike and railroad from the works will shortly pour in upon us a flood of Irishmen, who have signalized themselves on the Canal and elsewhere, . . . the advantage of the presence of arms seems now important."[23]

---

reports on the Chesapeake and Ohio canal and the prospects of navigation on the South Branch of the Potomac River. Virginia Board of Public Works, *Sixteenth Annual Report* (Richmond, 1832), pp. 375, 409; *Seventeenth Annual Report* (Richmond, 1833), pp. 77–93; *Eighteenth Annual Report* (Richmond, 1834), pp. 230–234; *Nineteenth Annual Report* (Richmond, 1835), pp. 414–431; Virginia Board of Public Works, Minutes, Mar. 26 and Apr. 4, 1833; Mar. 26, 1834; and July 1 and 13, 1836. Archives Division, State Library, Richmond, Va.

[23] J. H. Alexander to John N. Watkins, adjutant general, Oct. 1, 1838. Box 83, Adjutant General's Papers, Maryland Hall of Records.

There had been a good deal of labor trouble on the canal in 1838, and the adjutant general had already ordered general O. H. Williams of the Maryland militia to be ready for service "at a moments warning" with two hundred militia from the various regiments in Allegany County.[24]

However, the matter of lending arms to a private corporation seems to have been outside the jurisdiction of the adjutant general, for on January 3, 1839, we find William Matthews, in his capacity as state senator, submitting a resolution to authorize the governor to deliver to the George's Creek Coal and Iron Company one hundred stand of muskets "with their accoutrements complete." The resolution was voted down by the senate.[25]

October 2.—Additional article in relation to bailiffs added to the regulations. William Farley is retained, but Owen Richards sent off positively, having been drunk on Monday night again. Although he was in debt $7.50, J.H.A. directed him to clear himself [off]. Edmund Watkins is also discharged for drunkenness. Yonger engaged today at the setting and cementing of the south hot air furnace. Masons finish the inwalls at top and are engaged in filling up the foot-lock holes and plastering it with clay as they go down. Carpenter makes the rattles for the watchmen today, and blacksmith directed about the pontoons [spontoons—short spears]. The child of Edward Lewis, miner, is today badly burnt with powder which he had foolishly set fire to in a canister.

October 3.—P.T.T. and family leave today for Baltimore at 6½ o'clock A.M. Preparations made for commencing a new adit [drift], E, into the No. 12 ores. Bricklayers at the north hot air furnace, and the pipes all put up on the south one and 12 of the lower joints cemented. Our men can cement 8 joints apiece per day, but Yonger estimates that an expert hand could do 12.

October 4.—Dilly's contract at the top house yard is visited. Dilly assures me that his carts cost him $2 per day and that one cart answers for 8 men and is therefore for each man 25 cents. Also that he cannot get hands under $1.12½, making in all upon 5 yards [considered a day's work for one man] a cost of $1.62½, or for one yard 32½ cents. [This assumes that each man would dull 4 picks and that the cost of sharpening them would be 25 cents.] After dinner he was offered 35 cents per yard and no more. [Pauer's] family increased by a daughter.

October 5.—Plastering of the inwalls finished, and the hands are set to dressing bosh stones. Masons finish the store foundation today and the east wing wall of the dam, and are directed to commence with the sawmill foundation tomorrow. Koenig is positively ordered to leave the grounds tomorrow.

October 6.—A work [disturbance] is kicked up this morning about sundry boarders of Koenig having received no breakfast there. One of them who has the impertinence to come up to the residency, is discharged for that and for having been drunk the night before. Subsequently he is noisy and turbulent in Koenig's house and is handed over to the bailiff, who carries him off from the plantation.

October 9.—The case of John Williams, whose sons were detected in taking away plank, by the watchman, is considered, and John Hopkins directed to tell him that it will be passed by this time, but that if it occurs again he shall be discharged and severely punished. The case of Rees Rees, blacksmith, for drunkenness is also settled by his discharge, which takes place tomorrow. Letter from William Matthews received today, giving information of the consent of the Maryland Mining Company to the passage of the turnpike.

October 10.—The rain which commenced last night at 10 o'clock continues with great steadiness through the day. The engineer is about the steam engine, which is expected to suffer sadly. Orders are given to Mr. Welch [carpenter] to make a shed for it out of plank hitherto reserved for the store flooring. The hands are set about the furnace to cover it and save the plastering.

October 11.—Engineer engaged in leading the air pipes [using a lead compound to seal the joints] in the back alley. Finding, however, that it takes 65 pounds per joint, he is directed to use more spunyarn and less lead. Mr. Hopkins and family arrive today.[26] Mr. Serpell is constituted the superintendent of the turnpike. J.H.A. receives bad news from home, which seems to him to require him to absent himself after tomorrow morning.

October 12–18.—No records made in consequence of the absence of J.H.A. and P.T.T.

Tyson resumes journal:

October 20.—The boiler stack is finished at noon, and the bricklayers again work at the south hot air furnace. Ledley finishes burning his last kiln of bricks and departs with part of his hands. Nothing doing to the pipes for want of borings for cement.[27]

October 22.—At night a letter from Mr. Matthews about the jury for condemning right of way for the turnpike.[28]

October 23.—A fresh supply of borings have arrived, and the pipes of the northern hot air furnace are being cemented. Two contractors come to talk about the turnpike. [For the next six days Tyson was busy interviewing contractors and forwarding proposals to J.H.A. at Baltimore.]

October 24.—The hearth is lined with brick, and drying furnace erected at the tymp to dry the hearth and interior of the furnace. Fire made in it. [A new furnace had to be dried for as much as 8 or 10 weeks. A lining of common brick in the hearth prevented damage to the hearthstones during the drying process. The bricks were removed when the furnace was ready for charging.[29]]

[24] Orders dated June 21, 1838, published in Hagerstown [Md.] Mail, Oct. 4, 1838; and Walter S. Sanderlin, The Great National Project (Baltimore, 1946), pp. 121–122.

[25] Journal of Proceedings of the Senate of Maryland, Dec. Sess. 1838, pp. 11–12, 32, 35.

[26] The ship Harriet & Jesse arrived at Baltimore on Oct. 9 from Newport, Wales, with David Hopkins's wife and four children aboard. The 10 other passengers included 3 miners, each accompanied by wife and child, and one single miner. Although only one of these miners is mentioned in the journal, it is probable that the others also went to Lonaconing. Like the Tiberias, the Harriet & Jesse brought a cargo of railroad iron. Passenger list, National Archives, and Baltimore American, Oct. 8 and 10, 1838.

[27] "Iron borings are very scarce. I have caused the place [Baltimore] to be hunted over for them several times." W. Alexander to R. Graham, Oct. 16, 1838, Letter Book 1, Welch and Alexander Record Books.

[28] The act incorporating GCC&I Co. set out the procedure to be followed in negotiating for lands which the company would require for roads and railroads. If buyer and seller could not agree on the price, a jury summoned by the county sheriff would determine the "damages" which the owner would sustain by the "taking" of his property. Md. Laws 1835, ch. 328, sec. 8.

[29] Overman, Manufacture, p. 164.

October 25.—A letter from Major Powell, announcing that at a meeting of the citizens of Frostburg of which he was chairman it was resolved that they would assure the right of way for the northeastern 5 miles of the turnpike and pay $1,500 to $2,000 if we would terminate the road at that town.

October 30.—The steam engine set at work and continued for about 3 hours. It promises to do well. Letter to J.H.A. requesting him to get a casting for introducing air into the back tuyere.

October 31.—The bed of limestone that has cost so much in the search [see entry for June 30, 1838] is at length met with about 230 feet above the bottom of the big coal. Yonger and his assistants are adjusting the air pipes between the engine and the hot air stoves. Pagenhart's shantee takes fire, which is put out after producing a good deal of noise and some smoke.

November 1.—Messrs. Preuss and Pauer commence their levelings for the railroad.

November 6.—The furnace hands place the tuyere blast pipes. Arner fixes up the weighing machine near the foot of the inclined plane. Hopkins is having ore cleaned for roasting. The miners at drift E at a stand on account of rate of pay. Examined into the subject with John Hopkins and determined that they ought to receive $5 per yard [forward] and $3 per ton for the ore, and directed him to tell them to take it or give up the job.

November 7.—The engine is now entirely finished. The air pipes, however, are not all cemented. A letter from J.H.A. by Mr. Harrison, an additional engineer sent up by him for the railroad surveys.

November 9.—Weather very cold and blustering all day. Snow squalls all the morning. Bricklayers show symptoms of emigrating, but are induced to promise to work until 1st *proximo*. The weather being too cold to lay brick, they will work at the dam piers to get them out of danger. The weighing machine is completed, and the tramroad to it. Miners refuse to work at E level for $6 a yard and $3 for ore per ton.

November 10.—E level still quiet, the recusant miners getting out ore No. 14 from the patch. The weights for the ore weighing machine examined, and some of them marked. The ton of ore to consist of 21 cwt., or 2,352 pounds, so as to allow for waste, this being the Welsh custom.[30] [The weights were as follows: $\frac{1}{4}$ cwt. = $28\frac{3}{4}$ lbs.; $\frac{1}{2}$ cwt. = $57\frac{1}{2}$ lbs.; 1 cwt. = 115 lbs.; 3 cwt. = 345 lbs.] Welch lays wooden troughs for water to the engine.

November 13.—William Mitchell, a miner, ordered off for threatening violence to those who wish to take the work in E level.

November 14.—One of the miners (Harris the preacher) informs Mr. Hopkins senior that there would be a "row" because the men do not get $2 per day for day's work. Ordered both bailiffs to watch tonight. Bricklayers commence laying brick at the store. Finished weighing a lot of 20 cubic yards of ore No. 4. The ton is settled to be 2,415 pounds = 7 of the largest size weights. Mr. Serpell writes from Frostburg that the turnpike jury has awarded as follows: to William Wright, $97; H. Ross, $300; G. Stoup, $273; Winters, $375; total, $1,046.

---

[30] The Welsh and Cornish mining ton of 2,352 lbs. consisted of 21 cwt. of 112 lbs. each. As we see below, GCC&I Co. used a cwt. of 115 lbs. Weighing practices varied greatly. Alexander reports that in 1839 in Wales a "ton" of coal, ore, or fireclay weighed about 2,500 lbs. On the other hand, a "ton" of limestone weighed about 2,350 lbs. J. H. Alexander, *Manufacture of Iron*, p. 140. The weights described in the text were manufactured by J. Barker & Son, Baltimore. See Acct. Book entry Jan. 24, 1838, Welch and Alexander Record Books.

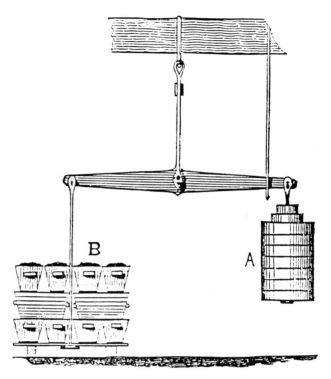

FIG. 7. Weighing ore. Overman, *Treatise on Metallurgy.*

Since the company professed to believe that it was benefiting its neighbors by making a good road where none at all had existed, the unwillingness of the landholders to donate rights of way for the turnpike was "unexpected" and "unnatural." Therefore, it announced, "it seemed proper to suspend all proceedings in regard to the letting of the road; nor will it ever again be taken up until it can be made free of land damages." [31]

November 17.—Weather clear and cold this morning, and as the mortar freezes, the brickwork of the store is suspended.

November 19.—The stock of ore stacked up is 1,119 tons. [Here a table sets out the number of cubic feet from each of the ore measures. The number of cubic feet per ton varies from $21\frac{1}{4}$ to 27, depending upon the type of ore.] Welch is directed to shed over the southwestern wing [of the store] for temporary storage.

November 20.—William Williams 4th miner discharged for getting drunk and beating his wife. Other cases of drunkenness among the Welsh, but the parties not identified. Some not at work are directed to be looked after [searched for].

November 21.—Five miners discharged for drunkenness and disorderly behavior. T. Lewis, keeper per *Tiberias*, fighting last night and drunk today. The water for the engine is let in from the dam and through the pipes to the covered reservoir.

November 22.—Masons get on well with sawmill walls. Duvall commences putting in the arms for the water wheel. The engine reservoir leaks and is being planked and puddled. Miners are tolerably quiet. Tom Lewis, the keeper, is still drunk.

---

[31] GCC&I Co. Report 1839, p. 7.

November 23.—About 40 miners who are now working in ore No. 4 refuse to work without a change in price. They are offered $3 per ton and told that *no more* will be given. Six of them are disposed to emigrate, and Mr. Graham commences to prepare to settle with them if he can borrow the money in the city [Lonaconing]. The rest of them have not yet decided what they will do. The water is again let into the boiler reservoir, which appears to be tight. Water is pumped into the boilers ready for starting the engine again tomorrow.

November 24.—Eleven of the malcontent miners are paid off and have permission to depart. Mr. Graham has great difficulty in borrowing money enough from the hands, the remittance from Baltimore not having arrived. The steam engine is put to work and does very well, but the air regulator is not sufficiently strong.

November 26.—Welch and his men continue to work at the cabins, which are nearly all finished. Arner prepares timbers for strengthening the regulator. Four men have commenced *E* level. Those who worked in No. 4 still idle. Rupert returns from Frostburg and reports that the eastern mail arrived there at 12 last night, but the deputy postmaster refuses or neglects to get up and receive it, so that our mail has journeyed to Uniontown.

November 27.—Mr. Alexander arrives from Baltimore.

Alexander resumes journal:

November 28.—The last day of grace afforded to the miners, the superintendent having given notice that all not at work in the morning would be discharged. Engine being thawed, and carpenters busy in erecting, or rather, continuing the shed over the engine.

November 29.—Mines as usual, save the discharge of divers miners who were refractory.

November 30.—Engine blowing [air through the blast pipes] for a short time.

December 1.—A general visitation made of the underground works by J.H.A. and P.T.T. Mr. Hopkins reports that the miners are ready to go to work in the thirlings [cutting passages] of No. 4, *C* level, at $1 per yard, where before they received $1.50 per yard.

December 2, Sunday.—Mr. C. B. Shaw arrives from Frostburg, where he had arrived the day before. [Shaw was the new superintendent. invited on September 27 to come to Lonaconing. The journal gives no clue to the reason for Tyson's leaving.] Mr. Matthews dines here and stays all night.

December 3.—Mr. Matthews goes to Cumberland in the afternoon bearing a letter to T. S. Alexander on the subject of the injunction to the tax collector.[32]

December 4.—August [Weisskettle], John [Wortman], and Schwenner are employed in building the piers of the blast pipes. Yonger cleaning the boilers preparatory to blowing, as is expected, tomorrow. Mr. Shaw is being inducted into the necessary local information.

December 7.—Engine commences to blow today, but the regulator is found to change shape so much as to be very unsteady and to put in peril the pipes. [A sketch at this point shows the distortion of the regulator, and the text describes the steps taken to remedy the difficulty.]

The blast at Lonaconing was furnished by a 60 horsepower engine having a steam cylinder 18 inches in diameter and 8 feet long. Steam was generated in 5 boilers, each 24 feet long and 36 inches in diameter, at a pressure of 50 pounds per square inch. The

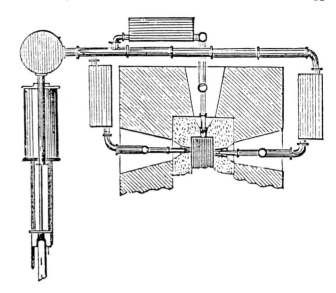

Fig. 8. Blast machinery and hot air stoves. The Lonaconing furnace blew through only 2 tuyeres and had only 2 stoves. Overman, *Treatise on Metallurgy.*

blast cylinder was 5 feet in diameter and 8 feet long. A piston rod extended through both cylinders. This apparatus forced into the furnace from 3,200 to 3,700 cubic feet of air per minute at an average pressure of $2\frac{1}{4}$ pounds per square inch. The back and forward movement of the piston produced an irregular blast, and it was necessary to pass the air from the blast cylinder through a large iron reservoir called a regulator, which equalized the blast and insured a uniform stream of air to the tuyeres.[33]

December 7 (cont.)—Wine is given to the miners upon the occasion of the installation of a new superintendent.

December 8.—[Tyson leaves. Alexander and Shaw accompany him to Frostburg.]

December 11.—At 1 P.M., J.H.A., C.B.S., and family reach Lonaconing from Frostburg. Afternoon spent in looking about and in hearing sundry complaints of a disturbance which had taken place on the preceding day. Letters received from T. S. Alexander enclosing the copy of the injunction bill and bond.

In 1838 the county commissioners of Allegany County reassessed the property of the George's Creek Coal and Iron Company, raising the value for tax purposes from $8,341.50 to $109,714. Since it appeared that the commissioners had not reassessed the property of other landholders in the vicinity, the company asked the chancery court for an injunction restraining the commissioners from levying taxes according to the new valuation. The court granted a permanent injunction in July 1839.[34]

---

[32] The company was planning to take action to protest the increase in its taxes. See entry for December 11.

[33] Alexander, *Manufacture of Iron*, pp. 92–93; and Walter R. Johnson, *Notes on the Use of Anthracite*, footnote p. 8.

[34] Chancery Court records No. 8282, Md. Hall of Records; Commissioners' Record, years 1813–1839, entry for Apr. 6. 1838; and Assessors' Records, Dist. No. 4, 1833–1841. The last two records are to be found in the basement of the courthouse, Cumberland, Md.

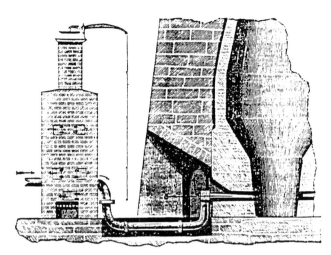

FIG. 9. Hot-air furnace showing blast pipe and tuyere entering blast furnace. Overman, *Treatise on Metallurgy.*

December 12.—After breakfast an investigation of the case of John Williams, one of the *Tiberias* miners, is held, from which it appears that he had occupied a house which had been positively refused to him by Mr. Hopkins. Upon the arrival of four Scotchmen who were placed in that house, John Williams was guilty of three times throwing their effects out. He also resisted and threatened the bailiff, who was sent to see them peaceably and properly installed. He was upon consideration discharged and ordered to clear himself and family by 10 A.M. on the morrow. The original contract, with an account, is sent by express to Cumberland in order to procure an attachment.[35] A case is also investigated between Blatter and Methusalem Rowland, from which it appears that there is no cause of complaint between them, but two boarders of M.R.'s are implicated. Their cases are postponed for a day or so. The regulator is finished, and under a pressure of 3 pounds per square inch yields hardly perceptibly. The regularity of the fly wheel is, however, not quite attained.[36] Schwenner and John Wortman are employed on the parapet [of the furnace stack]. Directions are given to August [Weisskettle] to have metal crampons [bars to hold the stones firmly together] put upon the upper courses of the parapet before the coping goes on. Double watch [by bailiffs] tonight.

December 13.—The regulator is found to stand a pressure of 3 pounds very well, but with some palpitations of the upper surface. The pressure was therefore directed to be reduced to 2 pounds for the constant working of the engine. The miner [Williams] mentioned yesterday comes this morning to ask pardon and to propose that his family be permitted to remain. Rupert returns with the message that the deputy sheriff will be here tomorrow, and the final determination of Williams' matter is left until that time. The hot air furnaces are tried, but an inequality of temperature between the extreme and middle vertical pipes is observed, the former being at bright red while the latter are not at all red. This is presumed to

arise on the one side from the free admission of oxygen through the manhole, and on the other from some access of air by the jointed entrance of the blast pipe. Mr. Hopkins is therefore directed to cool off the furnaces that tomorrow they may be entered and the leaky places examined.

December 14.—Hot air furnaces cooled off and entered, but no leaks discovered yet. At noon C. B. Shaw and J.H.A. go with Walters and [Michael] Milholland along the lumber road to inspect it. The determination is finally made to give $350 per mile and no more.

December 15—The bailiffs report a disturbance last night by one Meyrick Williams, miner, who is examined, acknowledges having been drunk, and is discharged. John Williams' case is also investigated, the sheriff having at length made his appearance, and he is finally discharged upon a written agreement to deduct from the wages of his boys $40 per month. J.H.A. becomes his surety for the repayment of sundry sums due in Frostburg, etc. He is permitted to remain until Monday morning with his family.[37] The peace of the community is profaned by what seems to have been a prize fight between Meyrick Williams aforesaid and one Thomas Abbott. Both are arrested and put under [the company's] militia guard until news is received from Elisha Coombes [magistrate] that he is too sick to appear tonight. Abbott is then sent home, and M. Williams held and put in the engine house for safekeeping under a guard of John Jones, William Hopkins, Rees Jones, and Richard Williams.

December 16, Sunday.—Meyrick Williams is called up, and an inquest held over him and those who abetted in the skirmish. From some want of intelligence or want of discretion of John Hopkins, to whom the matter of taking M.W. to Frostburg was given in charge, Meyrick refused to go. He was therefore put under guard, and as there was some riotous demonstration, an armed patrol was organized [in four shifts, two of nine men, and two of seven]. C.B.S. makes a grand round at 12 midnight.

December 17.—Rupert is sent at 6 A.M. to [magistrate] Billy Shaw, who arrives about ½ past nine. Meyrick Williams in the meantime placed under an armed guard until his case is tried. He is required to find bail to the amount of $100, which is done, and the money deposited with the magistrate. John Williams' case is again presented, and a supplementary agreement is proposed to be made tomorrow. Orders given to construct a flue or chimney in the center of the hot air stove in order to correct the false draught.

C. B. Shaw takes over the journal:

December 19.—Occupied in instructions from J.H.A. to C.B.S. and preparations for the departure of the former to Baltimore. Models of the furnace hearth explained to C.B.S. and deposited with him with instructions to ascertain their dimensions and give corresponding instructions to Weisskettle, with whom a contract has been made by J.H.A. to get out and prepare a duplicate of the furnace hearth [stones] for the sum of $500.

December 20.—Mr. Alexander departs on foot to Frostburg accompanied by C.B.S. the first 2½ miles. Milholland and Walters accompany us as far as Buskirk's with the purpose of examining a way to avoid by a new road the two-foot fording of the creek above Lonaconing. Mr. Alexander observes that Buskirk has been cutting timber near the boundaries of Skeleton Tract [Buskirk's land,

---

[35] The account showed that Williams still owed the company $187.62 for bringing him and his family from Wales. In addition he owed $47.07 at the company store. Clerk's Docket to April Court 1839, originals. No. 60, courthouse, Cumberland, Md. The contract could not be found with the other records.

[36] The number of revolutions per minute was 12. Johnson, *Notes on the Use of Anthracite,* footnote, p. 8.

[37] John Williams, one of the *Tiberias* men, had brought his wife and 6 children from Wales. The three sons described on the passenger list as "colliers" were aged eleven, thirteen, and fifteen.

bordering Commonwealth], and directs that a survey of the said tract be made by [Edward] Mullen to ascertain whether or not a trespass had been committed. [Mullen was at Lonaconing to resurvey Commonwealth in preparation for a suit against Buskirk to settle boundaries.] Complaint made by Mr. Graham that Mr. Coombes had been treating with wine some of the workmen who had been dealing with him, and that several of them were intoxicated, among them William Howell, the bailiff. On inquiry, find Howell much under the influence of wine, though at his work assisting George Blatter to build a watch house and belfry. Break Howell of his office and appoint in his place Rees Jones. Directions also given to Jones and Blatter to detail 4 men (2 Welsh and 2 Germans) to form a patrol for the night.

December 21.—A visit from Mr. [Israel] Mayberry, the tax collector, who proposes an execution upon the moveables of the company [for nonpayment of taxes], and that the superintendent should replevy, which course he finally abandons as a wanton tax upon enterprise and unnecessary injury to the company's credit, and merely takes an inventory of articles which are committed to the custody of C.B.S. pending the success of the bill which had been filed for an injunction. The articles listed are: 10 horses, 1 wagon, 1 transit theodolite, 1 Y level, and 1 Troughton's level.

December 22.—Find my kitchen not suitable for winter, and employ Weisskettle to line with brick. Visit the hot air furnace [and] find the new chimney to improve both the draught and heating power of the furnace, but not in a sufficient degree. [Here Shaw describes proposed changes in the hot air furnace.]

December 24.—Permission given to Mr. Welch to line with brick his dining room, it appearing that his dwelling would probably remain a portion of the proposed new town. Probable number of bricks, 2,500. Mr. Welch has several boarders. Direct Mr. Zacharias to give a pint of wine per hand to every man in the company's employment as a Xmas *douceur*, and to sell no drinkables on the 25th. Bailiffs sent round to give notice that strict compliance would be required with the ordinance respecting gun firing. The watch reinforced by two supernumeraries in each patrol.

December 25, Christmas.—Watch report all quiet the preceding night. Store closed, and a general holiday observed by the workmen. The day passes off with great quiet.

December 26.—Christian Blatter arrives from Pottsville bringing 2 German miners to examine and report to their countrymen the prospect at this work. Mr. Schmidt and David Hopkins directed to take all necessary measures to ensure as far as possible with economy the comfort of the different dwellings and to make such disposition of the quarters as may be necessary to accommodate the expected reinforcement from Pottsville. Mr. Mullen arrives in the evening very drunk, having appointed Thursday the 27th to commence surveys. C.B.S. determines to proceed with survey of Skeleton, but is disappointed by the sottish behavior of Mullen, who takes great offense at not being entertained (it is believed) by C.B.S.

December 27.—Mr. Mullen presents a bill for $21 and leaves Lonaconing in dudgeon. No loss, in the opinion of C.B.S., inasmuch as his habits require that he should be "dry nursed" in all his operations, and his services [would] consequently cost the company double at least those of another surveyor.

December 28.—Engine stopped for want of water. Dam examined for leaks, and the lower, or engine-house, dam found to be very defective. Germans employed in

repairs of dam. Bricklayers employed in laying grate for keeping warm the reservoir to steam engine.

December 29.—Lots drawn by Milholland and Walters for the two divisions of the timber road. Milholland obtains the first, or lower, division. Duvall has made some progress in the framing of the upper story of the sawmill. Mr. Graham complains that Duvall is drawing rather fast upon the company.

December 31.—Leak discovered in cistern at engine house [was] stopped, and working of engine resumed. Mr. Preuss and corps driven in by stress of weather. The junior engineers on leave of absence to Cumberland.

JANUARY 1–JUNE 30, 1839

The new superintendent's wide range of experience with the Virginia Board of Public Works qualified him to push forward the George's Creek company's proposed railroad and turnpike. In particular, his duties during 1835 and 1836 had brought him to the borders of western Maryland and into terrain which closely resembled that on the routes of the company's transportation projects. Moreover, whether or not Shaw had actual experience in mining, iron-making, and industrial construction, the journal indicates that he dealt successfully with problems in all of these engineering fields.

The combination of practical experience, theoretical knowledge, and technical ingenuity should have guaranteed that Shaw would be brilliantly successful at Lonaconing. Unfortunately, he was a very sick man. In his last report as principal engineer for the Virginia board, Shaw apologized for the delay caused by a "serious illness" from which he was just recovering.[1] Possibly this was the first manifestation of a chronic ailment. Shaw's symptoms, as noted from time to time in the journal, suggest that he had some form of cardiovascular problem, and that he frequently suffered from migraine headaches brought on by stress. Twice during the first six months of 1839 Shaw was almost completely incapacitated for long periods of time: the first from January 22 to March 2, and the second from April 27 to May 21. Consequently, Alexander had to be at Lonaconing a great deal more than he had expected (sixty-six days), and it was he who put affairs in train for the blowing in of the furnace.

Shaw was by no means as forceful, decisive, or ruthless as Alexander, who by virtue of his status as company president exercised an authority never fully transferred to his deputy. It was not until June 8 that Alexander gave Shaw "full discretion with regard to the general management of the works," and even after this date it is apparent that J.H.A. made important (and usually unpopular) decisions, leaving Shaw the thankless job of carrying out policy which he had no part in making. Shaw's role was

[1] Virginia Board of Public Works, *Twenty-first Annual Report* (Richmond, 1837), p. 374. The report was dated January 3, 1837.

made more difficult by his inability to deal easily with his subordinates. Sarcasm, moral outrage, and a sense of intellectual superiority punctuate his journal entries. Carried over to personal relationships, these qualities could not have failed to make his workmen rebellious and his contractors non-cooperative.

January 1, New Year.—Store closed and work suspended except in mines, where some work still goes on, but the product of the mines has been decreasing during holidays. Letter written to Mr. Oliphant, ordering castings for chimney dampers to the hot air stoves. The day celebrated by Mr. Graham in a dinner to the elite of Lonaconing—Mr. and Mrs. Preuss, Mr. and Mrs. Pauer, Mr., Mrs., and Miss Shaw, and Dr. Wundsch, guests. The evening spent most agreeably in music and waltzing.

January 2.—Information given to all the head workmen that hereafter a weekly report must be made of the work done in their several departments. An order dispatched by Mr. Graham for 6 new sets of tram wheels, axles, and boxes, two wheels having broken at the incline in consequence of the cold weather.

January 3.—John Hopkins returns from Oliphant's works. Oliphant's furnace out of blast. He promises J.H. to have our castings executed in the neighborhood. George Blatter, Rees Jones, and Rupert engaged in putting up a temporary market house.

January 4.—Letter from Mr. Oliphant. Castings will be ready by Saturday, 12th, but he seems to expect that we shall send for them. Reply, stating our inability to do so, and transmitting an order for iron and nails sufficient to make up the load.

January 5.—Mr. Schmidt by direction purchases a new horse for the mines—price $50—cheap. One more yet wanted. Petition (referred to Mr. J.H.A.) of 25 Presbyterians for the lease [of] a lot whereon to erect a church; also from Moses Ayres and others to use the permanent carpenters' shop for divine services. [The latter] granted, upon condition that a sufficient number of signatures be appended to the request. About 20 are handed in, which is thought to be sufficient.

January 7.—Give directions to take down the grate in the northeast hot air stove and remove it higher, with a view to bring the heat and flame nearer the pipes. Sommer employed in digging rough molding sand near smith's shop, and Richard Williams sent in search of some [sand] of a finer quality, which he discovers at the distance of about 6 miles. Re-laying of rails progressing in the C and D levels, and air pipes in the windways.

January 9.—Letter written to president of company [J.H.A.], communicating information concerning the works. Whole number of miners and colliers, including boys and laborers on the tips, 126. Number of tons raised in the month of December, 266 tons, 17 cwt. ore. No account of coal taken

January 11.—Mr. Serpell establishes the four corners [of] superintendent's house.[2] [Receive a] letter from Oliphant's works giving information that the castings

and flour expected from him would start from Uniontown on Monday, 14th.[3] The new bailiff arrives.

January 12.—Some dissatisfaction expressed by the colliers with their wages, and some appearance of a strike among them. Mr. [Joshua] Hill, the new bailiff, employed in obtaining a knowledge of the localities under the direction of Rees Jones, whom he is to succeed. George Blatter employed at top house yard preparing clamps to burn "mine" [ore]. At 6 P.M. Bible Society meets. Attendance not as good as might be wished, none of the young gentlemen, though all were in the town. Thirty-seven members subscribed $2 per annum. Whole amount of subscriptions and donations about $100. Order taken for purchase of Bibles, and the meeting addressed by C.B.S.

January 13, Sunday.—Visit from Mr. Howell and Messrs. Young of the West Point Foundry. [William Young later left the West Point establishment to become president and general superintendent of the Maryland and New York Coal and Iron Company at Mount Savage.]

January 14.—Two wagons arrive with 22 barrels flour from Pennsylvania, purchased by Mr. Graham at $7.87 and $7. Accounts of McVickar, the butcher, examined, and a balance of $8 or $9 found in favor of the company upon two weeks' occupation of a butcher. He is directed to continue the purchase and slaughter of cattle and to make a slaughteryard.

January 15.—Keepers at furnace engaged in fitting universal joint to nozzles of blast pipes. Snow all day without intermission. New bailiff directed to add a room to the watch house for his own sleeping accommodation.

January 17.—Settlement day. A riot occurs. David Lloyd detained all night in the new watch house. A strong disposition evinced on the part of the miners to rescue Lloyd. They are threatened with dismission in the order in which their names may be recorded on the spot. Patrol of Germans at night. Lloyd furious and endeavoring to break prison, but becomes calm in the course of the night.

January 18.—Messrs. William Shaw and Elisha Coombes hold an inquest into proceedings of previous day. Lloyd dismissed for assaulting bailiff when in his custody. A riot at night, and two men (Morgan Howells and Daniel Griffiths) committed to "quod" by Mr. Hill, the bailiff, assisted by Richard Williams and George Blatter. D. Griffiths minus his under lip, and M. Howells much mutilated with bites about the face and head.

January 19.—Riot of preceding night investigated and found to have proceeded from drunkenness, Christopher Howell principal instigator and Mrs. Harris (widow) accessory. C. Howell dismissed, and, in company with Lloyd, escorted off the company's land at 3 P.M. by the bailiff. Mrs. Harris permitted to remain until Monday to sell her effects. Morgan Howells in danger from his injuries. Meeting of Bible Society postponed on account of the disturbed state of the community. The miners and colliers, however, who had on the preceding day refused unanimously to work, return to their duty today, having been threatened with dismissal by alphabet at the rate of 5 men per day, and the names of the first 5 made known the night previous. They seem now to be better satisfied that the charter and by-laws of the company confer jurisdiction, in cases of misdemeanor, on the superintendent, and that the bailiff is a legal officer, the rules and regulations having been read and translated

² This house was in the vicinity of the new store and new sawmill near the junction of Koontz Run and George's Creek. R. Graham to R. Wilson, Oct. 23, 1839, GCC&I Co. Letter Book, Md. Hist. Soc. It apparently never served as the superintendent's house. Instead, the "stone house" occupied by Graham became the superintendent's residence during Graham's caretaker superintendency (1840–1849). After the Civil War, two American Gothic wings were added, the grounds were fenced, and the estate was named "Alexander Park." The older inhabitants of Lonaconing remember it by this name.

³ In hard times Oliphant would sell iron to farmers in his vicinity and accept payment in kind. From time to time he loaded his wagons with flour and other goods to be sold in the towns along the National Road. Franklin Ellis, ed., *History of Fayette County, Pa.* (Philadelphia, 1882), p. 584.

to them and the opinions of the magistrates, Messrs. Shaw and Coombes, made known through Mr. D. Hopkins.

January 20, Sunday.—The miners and colliers addressed at 3 P.M. by the superintendent on the subject of temperance and orderly conduct, and Monday evening appointed for the formation of a temperance society.

January 21.—Experiment tried with northeast hot air stove after raising fire grate. The success beyond expectation. The pipes are now equally heated, and the temperature of the blast is raised to the melting point of lead [about 622°F.] without difficulty. Directions are given to Weisskettle for alteration of southwest stove. Temperance meeting at 7 P.M. Society organized. C.B.S., president, and Richard Serpell, secretary. Seven members on principle of total abstinence, and 26 on that of abstinence from ardent spirits. Monthly meetings to take place on the third Monday of each month.

January 22.—C.B.S. discovers that he had sustained such injury from exertion during the difficulties of last week as required medical aid. Advised by the doctor to keep at home for the present. Mr. Serpell sent for and instructed in the various duties of the superintendent, which he is required temporarily to discharge.

January 23.—Thomas Layson, miner, has his leg broke by a fall of ore from the roof of the mine. Fracture compound, but the doctor pronounces the case manageable and that he will be well in 6 or 8 weeks.

January 24.—Directions given to Clark, blacksmith, to fit the levers etc. for working the new chimney dampers to stoves at furnace. Mr. Harrison returns a copy of a drawing which he had been directed to make of "Palmers Balance Railway." [4]

January 25.—Rosenbauer, miner, has his thigh bone fractured in the C level and is otherwise dangerously hurt. A messenger and letter dispatched to Drs. James and Samuel Smith of Cumberland to consult with Dr. Wundsch in this desperate case. Doctors arrive from Cumberland at 9 P.M. and confirm the opinion of Dr. Wundsch, previously expressed, that nothing could be done for Rosenbauer, the fracture of the thigh being too near the hip for amputation; in addition whereto, the internal injury seems itself to threaten dissolution, no reaction having followed the accident, the pulse feeble, and evident congestion of the lungs and lower viscera.

January 26.—Rosenbauer still survives, but without the slightest hope. The physicians depart for Cumberland, having been paid their fee ($20). Rosenbauer lingers until midnight, and expires. Mr. Ledley, brickmaker, and three journeymen arrive this day from Baltimore.

January 27, Sunday.—Drifting snow all day. Orders given to Mr. Welch to make a coffin for Rosenbauer, who is to be interred on Monday in Mr. William Shaw's cemetery. Rupprecht finds the snow in some parts of the road to Frostburg 7 feet deep.

January 28.—No work done this day, being the one appointed for the funeral, which takes place at 2 o'clock. Funeral service performed by the Reverend Mr. [blank], German preacher from Cumberland.[5] Fire breaks out in house No. 57 occupied by John Lewis, keeper. That and No. 56 almost totally consumed. No. 56 tenanted by Dan Simmons, fireman.

January 29.—Orders given to Mr. Welch to hang doors and window shutters upon the old logs of No. 56 and No. 57, and to Mr. Schmidt and corps to repair damages with all dispatch. The families of Lewis and Simmons temporarily provided for.

January 31.—Steam engine discovered to be slightly out of order—a leak in the feed pipe, which "sucks air." J. Jones employed in repairs. Paving of the bridge house floor, which had been begun by [Mathias] Gruber, discontinued.

February 2.—C.B.S. entrusts himself again to the perils of outdoor locomotion, and visits the hot air stove, furnace, smith's and carpenter's shops, and the office. Things seem to be going on as usual, except that the constant demand for smith's labor has rather delayed the fitting of the chimney dampers to [hot air] stove.

February 3, Sunday.—Service in Mr. Shaw's chapel by the Reverend D. Jones, one of the miners—a very sensible discourse delivered partly in Welsh and partly in English.

February 4.—C.B.S. again visits the works, but in much pain. McVickar, the butcher and drover, returns from an unsuccessful trip to Virginia in quest of cattle and sheep. A new route is indicated to him, and as he may have far to go, it is determined after due consideration to authorize the purchase of 4 head of cattle. All kinds of provisions becoming scarce in the neighborhood. Most fortunately, Mr. Graham's Wheeling correspondent has announced as on the way a load of bacon, cost price 9½ cents the pound. Sundry accidents to the miners in the course of the last week—generally slight ones. Dr. Wundsch very busy.

February 5.—Directions to Arner to make new furniture for the burnt houses. One of the boilers discovered to be scaling off next to the furnace. Must have been made of cinder iron [inferior grade], cannot last long, and is undoubtedly dangerous. Mr. Alexander to be consulted. Complaint made by two of the German miners, Schmidt and Haass, that John Hopkins does not give them the same advantage in their work as is given to the Welsh miners. Mr. Alexander arrives at 2 P.M. accompanied by the Reverend Mr. Huntington, the chaplain.[6]

February 6.—Mr. Alexander writes to the mayor of Baltimore [Sheppard C. Leakin] for two pair handcuffs.

February 7.—Mr. Alexander inspects condition [of lumber road] as far as Milholland had progressed, C.B.S. not yet being sufficiently convalescent to proceed so far. At night Mr. A. determines the organization of the military company for the defense and protection of Lonaconing. Officers and non-commissioned officers: C. B. Shaw, commandant; F. Pauer, 1st lieutenant; F. Schmidt, 2nd lieutenant; John Hopkins, 3rd lieutenant; Richard Serpell, 1st sergeant; Harrison, 2nd sergeant; William Webster, 1st corporal; Richard Williams, 2nd corporal; privates [no list given].

February 8.—Drover from Virginia offers cattle averaging 6 cwt. 3 quarters at $65. Mr. Graham is directed to purchase 6 head. C.B.S. indisposed. Mr. D. Hopkins reports the complete success of the new chimney dampers. Thomas Layson's broken leg is still very bad. Dr.

---

[4] An early version of the monorail, illustrated in Henry R. Palmer, *Description of a Railway on a New Principle* (2d ed., rev., London, 1824). See also Thomas Tredgold, *A Practical Treatise on Rail-Roads and Carriages* (London, 1825), pp. 36–38.

[5] Probably the Reverend John Kehler, pastor of the Lutheran church in Cumberland from 1832 to 1841. J. Thomas Scharf, *History of Western Maryland* (Philadelphia, 1882) 2: p. 1479.

[6] Possibly the Reverend David Huntington (Protestant Episcopal), who had recently requested transfer from New York to the Diocese of Maryland. Benjamin T. Onderdonk, bishop of New York, to Standing Committee of Diocese of Maryland, Nov. 24, 1838. Md. Diocesan Archives, on deposit with Md. Hist. Soc. J.H.A. noted an advance of $50 to "D. Huntington" on Jan. 2, 1839, and expenses of himself and "D.H." to Lonaconing, $50, on Feb. 5, 1839. Vol. 1, Hammer Acct. Books, Md. Hist. Soc.

Wundsch very much occupied in administering to sundry wounds, bruises, and putrefying sores.

February 9.—C.B.S. sick and confined to bed. Arrangements made for Episcopal service in Lonaconing church. Organization of military company communicated to Messrs. Pauer, Schmidt, and Hopkins, and directions given them to ascertain whether the privates would equip themselves. Directions to J. Hill [bailiff] to ring the bell at sunrise and sunset of each day, at 12 and 1 o'clock P.M., and at 9 and 12 P.M. and 5 A.M. Church bell to commence at 10½ A.M. [and] to be rung until 11.

February 10, Sunday.—Morning service by Mr. Huntington.

February 11.—Snow at intervals through the day. Rupert does not go for the mail. Joshua Hill and Mr. W. Webster visit the various domiciles of the place to ascertain the probable number of scholars at the proposed school of Mr. Huntington. Directions given to them to take inventories of the hogs and dogs upon the place, with a view to enforce the existing ordinance in relation to them.

February 12.—Mr. Webster returns the number of children for school, 64: ditto too small, 94; the number of dogs kept on the place, 36; and the number of hogs not penned [not given]. Directions to be given the bailiff to abate the latter nuisances. J. Hopkins' account of the work of 3 German miners disputed by them and appeal taken to C.B.S.

February 13.—Measurement of the German miners' work by Mr. Serpell, shewing 75½ yards in the long work of the No. 12 ores to have been excavated in 6 weeks, which at $4 per yard makes $301.33, or $100.44 to each man. Their monthly wages were therefore $67 per man. [The cost of ore] may be estimated at $8 per ton. Determined to discontinue this working at the end of the current month in consequence of the expense.

February 14.—Wagon arrives from Baltimore with goods, and arms for the proposed volunteer company. It is discovered that the cliff above the patch had come down during the night, entirely burying the workings of the German and Scotch miners in the No. 12 ores. Fall computed to contain 1,500 tons of matter. Mining tools are lost, but will be recovered in the advance of the heading to left hand of D level.

February 15.—Select a site for the new Episcopal church, which is staked out by Mr. Preuss and the foundation leveled off by Mr. Schmidt's men the same day. At night estimate the cost of a log building 44 feet by 24 and 12-foot pitch, with a log partition 12 feet from the east end, having 16 pews and capable of containing 100 persons. Estimate $644.

February 16.—Select site of new tavern. Visit Dr. Wundsch's house, and direct Mr. Preuss to make out an addition 16 feet by 12. Directions given to Welch to get out the frame and commence the building forthwith. McVickar the butcher has returned from Virginia, having purchased 4 head of cattle on Cheat River of Jacob Lee and engaged 20 head more to be fed 6 weeks and paid for when taken away at $43 per head. Beef in market.

February 17, Sunday.—Morning service from Mr. Huntington. Afternoon service by Mr. Forrest from Frostburg.[7] At night find two miners. John Hughes and J. B. Jones, in custody of the bailiff. Thomas Lewis, keeper, also arrested and confined all night. Patrol.

February 18.—Mr. Pauer is dispatched to Frostburg for magistrate and constable. Appoints Mr. [John] Porter and E. Coombes, Esquire, to visit Lonaconing on the 19th. Night patrol increased to 16 men in consequence of sundry disorderly persons lurking about the place and holding a carouse at Buskirk's. Orders given to post 2 sentinels at the dam and 2 at Hill's Run to take down the names of all intoxicated workmen who may enter the place from the direction of Frostburg, and to convey the very disorderly ones to the guard house, where the 3 men already arrested still remain. Written orders also given to the officer of the guard to use all possible forbearance in case of any actual disorder, and not to use firearms unless the guard or guard house be attacked. Part of the muskets are loaded with small buckshot, and directions given that they shall be [the] first [to be] discharged, and with a low aim. A note is also dispatched to Mr. William Shaw, requesting his presence as a magistrate as soon as he can come.

February 19.—The night passes off quietly. The revelers at Buskirk's pass the sentinels by keeping high up the hill. Benjamin Thomas, miner, and wife stopped by sentinels in a state of intoxication. Mr. [William] Shaw cannot attend. Mr. Coombes comes in the afternoon and inquires into the difficulties of the previous two days. John B. Jones convicted of an aggravated assault upon Mr. D. Hopkins, and is required to give security in the sum of $10 to keep the peace and appear at the April court. Thomas Lewis, keeper, and John Hughes are discharged, and T. Lewis being unwilling to pay a balance of $36.67 due to the company, and Jones to find the requisite security for his good conduct, both are remanded to the guard house for the night. Patrol continued tonight.

February 20.—Porter, the constable, arrives from Frostburg and departs for E. Coombes' with Lewis and Jones, the former of whom pays up his debt to the company, and the latter deposits, on the part of John Jenkins, miner, the sum required as security for his good conduct. David Evans, miner, also dismissed for insulting behavior to the bailiff.

February 21.—Way reconnoitered for a cart road from limestone quarry to bridge house. The Germans, 9 in number, are not at work, and complain that they cannot obtain trams to carry out their rubbish. There seems to be little doubt that they are much annoyed by the Welsh miners. John Hopkins has refused them horses and laborers (tippers) for night work. He is sent for by C.B.S., but does not come. Mr. Ledley, brickmaker, arrives from Baltimore with more hands. Hot air stoves finished. Wortman engaged upon the chimney stacks to them.

February 22.—Nativity of Washington. Rupert goes to Frostburg in cart to bring Mr. Ledley's baggage and some books for Mr. Alexander.[8] Mr. Serpell makes an unfavorable report of the progress of Walters' division of the lumber road. Walters is sent for and reprimanded.

---

[7] The Reverend Josiah Forrest (1797–1873), in charge of the Frostburg circuit of the Methodist Episcopal church. Scharf, *Western Maryland* 2: p. 1479; and James Edward Armstrong, *History of the Old Baltimore Conference* (Baltimore, 1907), p. 375.

[8] The books were: William Fuller Pocock, *Architectural Designs for Rustic Cottages* (London, 1807); C. A. Busby, *A Series of Designs for Villas and Country Houses* (London, 1808); and John C. Loudon, *Encyclopedia of Cottage, Farm and Villa Architecture* (London, 1835). Both Loudon and Pocock devote considerable space to houses for laborers and mechanics. One surmises that Alexander saw himself in the role of the benevolent employer providing his workmen with "comfortable habitations, where, after the labours of the day they may enjoy domestic comforts in the midst of their families." Pocock, p. 5. The designs are decidedly picturesque and are more suited for the English countryside than for a mining town in a wilderness region of America.

Mr. Huntington dines at Mr. Graham's with Mr. Alexander. The health of Queen Victoria is drank by the company. J. Hopkins is directed to furnish the German miners facilities, and to let them work at night. Mr. Schmidt has made sundry horse trades which result in an increase of the cavalry force of two horses.

February 24, Sabbath.—Mr. Huntington preaches in the morning; Mr. Venable in the afternoon. Some disorderly persons detected firing guns near the town—W. Blutter (boy) and one of Duvall's men. Blutter's father promising to administer the necessary punishment, no further notice is taken of his offense.

February 25.—Visit the tips, which are still on fire. Some careless workmen have fired the outcrop of the 8-foot coal. J. Hopkins is directed to have it extinguished at any cost. J. Hopkins applies for increase of salary. Regulations read by C.B.S. to Duvall's and Ledley's men, who are informed that they must conform to all the rules of the residency. Job Davis is reported by the bailiff as disorderly. His case postponed. A. Funke also reported as selling liquor and permitting gambling in his house. He is sent for and admonished. Powder magazine reported leaky. Examined by Mr. Alexander, [and] a coat of pitch and gravel ordered.

February 26.—It is discovered that Hoffman, the smith, has a barrel of whisky in his house. The bailiff is directed to search his house and to give to Hoffman his option whether to convey himself and his liquor from the company's premises, or to stave his liquor. He prefers the latter alternative, and destroys the whisky, which amounts to within 2 gallons of a full barrel. C.B.S. (as usual) indisposed, but attending as much as possible to business.

February 27.—C.B.S. very much indisposed. Mr. Alexander attending to business. Contract for the new church at the mouth of Koontz Run given to Conrad Fazenbaker, who undertakes to build the shell of logs hewn on both sides, and to frame and shingle the roof, for $200. D. Hopkins and the furnace keepers engaged in filling the coke yard.

February 28.—Snow again all day. Visit from Walters, one of the contractors on the Savage road. He has been going on badly for some time. He is told that he cannot be continued in credit at the store, and that his immoral habits have deprived him of the company's favor, and that he will not be continued as contractor after this job is finished. Visit steam engine and examine into its condition. J. Jones ordered to repair the defective boiler. J. Jones applies for increase of salary. Mr. Alexander talks to him "like a Dutch uncle."

March 1, St. David's day.—Mr. Alexander departs in the sleigh for Frostburg and Baltimore. Holiday among the Welsh. Dinner by the Cymraeggyddion Society. Mr. Graham, Mr. Pauer, and Mr. Huntington, guests. C.B.S. [unwell] unable to attend. His apology and an address from Mr. Alexander read to the meeting by Mr. Graham. Healths drank—of Mr. Alexander, Mr. Graham, Dr. Macaulay, Mr. Wilson, and others connected with the company, its prosperity, and that of the temperance society—all in good feeling and perfect decorum.

March 2.—C.B.S. confined to house with gatherings in both legs, pronounced by the doctor to be rheumatism and prescribed for.

March 3, Sunday.—Mr. Huntington preaches to thin audience—no one from the stone house [Graham's]. Mr. Duvall arrives from Baltimore and brings some vaccine matter for the doctor.

March 4.—Moses [Ayres] has valued Spiker's house on the hill at $20. It is determined to dislodge him to make room for the German miners. Mr. Schmidt is also directed to contract, if it can be done for $45 each, for two more double cabins. Mr. Welch is directed to hasten the fitting up of Mr. Huntington's room in Rupert's house, and Mr. H. is invited to stay with C.B.S. until that can be done. Mr. Huntington is to commence the school on Wednesday.

March 5.—Henry Spiker applies for permission to remain longer in his house. He is only permitted to stay until Wednesday the 6th, and an order is given on Mr. Graham for the value of his house ($20). Spiker has been keeping a house of bad repute for some time. Hopkins has given notice to the miners in the No. 4 ores that they will be reduced to $3 the ton, and there seems to be an unwillingness to work by the ton unless the miners in the No. 12 are also rated by the ton. Hopkins finally compromises by reducing the price in the No. 4, and abandons the idea of putting the work by the ton for the present. The prices now paid are $3 per yard in the longwork of the No. 12; $1 per yard on the No. 4, the men to find their own powder and candles.

March 6.—Mr. Serpell is directed to visit Walters' section of Savage road and ascertain its condition and whether Walters' foreman, Johnson, will finish his contract. Walters is very ill with nervous fever. His contract is declared forfeit, and is taken by Johnson upon the estimate of Mr. Serpell. Milholland's division will require work (extra) upon a single difficult mile to the amount of $1 per rod. Mr. Johnson also finds rock upon one of his turns when the cut was deep, and estimates the increased cost at $130. Letter at night from Mr. Alexander, introducing John Miller [another junior engineer], who is put under the care of Mr. Serpell. Mr. Miller's wages to be $40 per month and a house.

March 7.—John Jones, employed in repairing steam engine boilers, finds them very defective. Rees Rees, blacksmith, dismissed for drunkenness.

March 9.—Mr. Alexander is informed of the defective state of the boilers, and it is suggested that no final settlement be made with the West Point Foundry until they are inspected. A wagon arrives from Baltimore with Bibles and prayer books and goods to store. Gaiters, belts, and swords for the military company also arrive in the wagon. Verbal contract with Mr. Welch for the carpentry of the store, the tavern, the top and molding houses, and the superintendent's house. He is to employ 10 men at once and 10 more at the requisition of the superintendent; the force is to be increased to 30 if required, and the contract is to be annulled if the work does not seem to be progressing with proper diligence.

March 11.—Mr. J. H. Alexander's letter of March 10 furnishes the prices paid at a charcoal furnace at Laurel Hill to be entered in the journal.[9] Mr. Brown of Nova Scotia arrives, bringing a letter from Mr. Alexander and $3,000 for the settlement. Mr. Brown visits the coal level, the furnace, and ore drifts, dines with C.B.S., and takes leave. The miners in the Sydney and Pictou coal mines raise per man from 10 to 15 tons per week. Mr. Brown thinks the wages allowed here very liberal.

---

[9] Possibly Fountain Furnace, at the foot of Laurel Hill, Donegal township, Pa. James M. Swank, History of the Manufacture of Iron in All Ages (2d ed., Philadelphia, 1892), p. 219. The National Road crossed Laurel Hill a few miles east of Uniontown. The wages paid, per ton of iron produced, were: ore diggers, $5.50; colliers [charcoal burners], $4.50; founder, 58 cents; 2 keepers, 58 cents; 2 fillers, 50 cents; 2 bankmen, 50 cents; 2 guttermen, 22 cents; hauling ore, $1.50; hauling coal [charcoal], $1.12½; clerk and manager, 75 cents; engineer and smith, 52 cents; total, $16.27½.

March 12.—Order to limit the price of board [meals] to $3.

March 13.—C.B.S. makes the quarterly report to the president of the company, wherein it appears that the yield of ore for February has been 408 tons, the number of miners and laborers connected with the mines being 140, and their wages $5,362, making the cost of ore, even under this very visible improvement, $13.14 per ton. The 241 tons of the month of January [cost] $4,792, or $19.88 per ton. Complaint is made by the Scotch miners of partiality on the part of J. Hopkins. They state their case, [which is] to be inquired into.

March 14.—John Hopkins is questioned concerning the complaints of the Scotch miners. He denies the truth of the allegations.

March 16.—Fazenbaker has finished church up to the square and applies for rafters and shingles to C.B.S. He is told that his contract, according to the record in the journal, required him to find them. Letter at night from Mr. Alexander, announcing the employment of W. Shaw at $800 per annum. Directions given as to his occupation.[10]

March 18.—W. Shaw arrives, bringing a letter from Mr. Alexander enclosing a petition against a reassessment of property in the county. The necessary steps taken to procure signatures. Eleven bricklayers arrive, sent by Harris. They refuse to work except by the day [that is, at a daily wage instead of a rate per thousand]. Harris' nephew will not contract without previous correspondence with his uncle. He is offered work for his hands by the day until Mr. Harris can be heard from.

March 19.—Hill, the bailiff, in pursuance of orders, kills a hog of Koenig's, which had long been trespassing. Much excitement produced among the Germans in consequence. Dittmer is insolent, and is dismissed. Dr. Wundsch espouses the national dispute, and is told that he is not wanted unless he chooses to obey the ordinances. His account is directed to be adjusted, and Dr. Hunt is sent for to attend until a new surgeon can be appointed. The bricklayers, except four, leave Lonaconing.

March 20.—Dr. Hunt arrives from Frostburg. Dr. Wundsch still contumacious. Miners refuse to work on account of Dr. W's dismission. They are told that they are all welcome to go, and that their wages must soon be reduced. They go to work in the evening.

March 21.—Mr. Huntington leaves Lonaconing. Dr. Wundsch shews signs of repentance. Dr. Hunt still attending. Examine new cabins, 3 in number, with Mr. Schmidt. Thomas Jones killed in coal level. Fazenbaker does nothing to the church.

March 22.—Letter to Mr. Alexander explaining the occasion of Mr. Huntington's leaving the place [never explained in the journal]. Meeting of the military company, and funeral of the miner who was yesterday killed in the coal level. No work today except among the laborers.

March 23.—Dr. Wundsch continues his attendance, but today for the first time shews a willingness to quit the field in favor of Dr. Hunt. Water let into the sawmill race, and the wheel tried at 6 P.M. Contract made with Mr. Conrad to come on Monday and bore pipes for the

water tuyeres at 6 cents per foot, himself and a hand to be boarded. Rupert is spoken to to take them.

The water tuyeres used at hot-blast furnaces were conical tubes of cast iron. To prevent their melting, they were made hollow, and a steady supply of cold water flowed through them. The cold water came in through one pipe, and the heated water was carried off by another.[11] The company planned to bring the cold water to the tuyeres through log pipes, small logs with a hole bored through them from end to end. The logs were preferably of white pine, about 6 to 10 feet long, and the hole was usually $1\frac{1}{8}''$ in diameter. One end of each log was shaped into a cone; the other end, into a cup. Thus the pipes could be joined fairly snugly by fitting cones to cups.[12]

March 24, Sunday.—Sermon from Mr. Richards of the Baptist church. C.B.S. writes to Dr. Wundsch, offering to reinstate him upon his submission.

March 25.—Contract made on Saturday [with two Welshmen] for quarrying limestone on the creek near Arner's at the rate of $2 for every 128 cube feet, the stone to be piled and then measured, and 20 cents per yard for stripping. The contractors to be found in tools for stripping and to be charged with the tools necessary for quarrying, their tools to be kept in order, and they to find their own powder.

March 26.—Dr. Wundsch makes a proper submission by letter, and Dr. Hunt's attendance is discontinued.

March 27.—J. Hopkins is directed to announce a reduction of $12\frac{1}{2}$ cents per yard in the stalls and 50 cents upon the road in the mines.

March 28.—A contract made with Broadwater to erect a log house for Mr. Hill near Mr. Pauer's garden for $50.

March 29, Good Friday.—Germans keeping holiday. Moses Ayres and Pagenhart engaged in flooring the new cabins and hanging doors. A contract is made with [Weisskettle] for the brickwork of the store at $3 per thousand, the company to find all materials. Mr. [David] Hopkins directed to make every preparation to commence the blast.

March 31, Easter Day.—Preaching from the Welsh preachers through the day.

April 1.—Holiday among Welsh and Germans.

April 2.—Hill, the bailiff, reports misconduct on the part of some of Duvall's hands the preceding night. Duvall is required to admonish his apprentices and inform them that they must comply with the regulations or take the consequences. Machinery for shingle and lathing machines in progress, the new sawmill in the meantime making an occasional cut with the water of Koontz Run. Conrad begins boring the tuyere pipes.

April 3.—Weisskettle directed to put a stone pier and abutment for a footbridge across the race opposite the new church. Broadwater commences Hill's house, and Hill sets his garden fence (near Pauer's). Miners apparently still satisfied with their wages; none have left but the Germans.

[10] The journal entry of April 9 sets out the terms of the commission of William Shaw, local landowner and magistrate, as company agent. The commission authorized him to supervise company lands not currently being developed for mines or other purposes, and to arrange for renting or leasing them. The company had about 65 tenants on small farms within the Commonwealth and Beatty's Plains boundaries. Testimony of Robert Graham, supt., in GCC&I Co. v. C. E. Detmold, Chancery Records No. 8284, Md. Hall of Records.

[11] Charles Tomlinson, ed., Cyclopedia of Useful Arts & Manufactures (London and New York, 1854) 2: p. 78; Frederick Overman, The Manufacture of Iron in All Its Various Branches (Philadelphia, 1850), pp. 418–420; and Overman, A Treatise on Metallurgy (6th ed., New York, 1882), pp. 415–416.

[12] Jared Van Wagenen, Jr., The Golden Age of Homespun (Ithaca, 1954), pp. 123–125.

April 4.—Arner is directed to take charge of the lumber and charge all deliveries to their proper accounts, for which purpose he is duly instructed.

April 6.—Four thousand feet [of] boards arrive from steam sawmill. One thousand given at once to Ledley to make a brick shed, and sundry smaller parcels given out for cabin floors. Contract with Gruber, Grauss, and Romeyser for brick work of superintendent's house at $40 each for the chimneys and $3 per thousand of bricks for the lining. Messrs. Alexander and Tyson arrive from Baltimore.

April 8.—Messrs. Alexander, Tyson, and Shaw visit after dinner the *C*, *D*, and *E* levels and the stalls on the outcrop of the No. 4. Messrs. A. and T. direct that the work shall all be squared up preparatory to putting it by the ton. The price [is] fixed at $3 the ton, to include the cutting of the road.[13]

April 9.—Mr. Serpell's family remove to Lonaconing, his house, with the exception of the kitchen, being completed. Clark directed to iron the garden roller, and Mr. Schmidt, after rolling the garden walk, to allow Mr. Ledley the use of it for leveling the brickyard, and when he [Ledley] is done, to let Mr. Hopkins have it for the coke yard. Mr. Schmidt directed to examine a wagon and team reported to be for sale near Cumberland. Computation of the cost of making iron at these works made in the evening by P.T.T., C.B.S., and J.H.A. [The computations are in J.H.A.'s handwriting.]

The computations are summarized to show daily expenses of the furnace, and the cost per ton of pig iron produced, assuming a cast of 10 tons per day and the lowest expected cost of materials.[14]

Wages per Month

David Hopkins	$112.50
John Hopkins	70.00
2 keepers @ $60	120.00
2 fillers @ $50	100.00
1 molder	40.00
engineer	60.00
D. Simmons, fireman	50.00
additional fireman	40.00
mine burner	40.00
bridge stocker	40.00
2 driving boys @ $15	30.00
contingencies	100.00
Total per month	$802.50
Total per day	26.80

[13] Oliphant's miners received $2.40 per ton, but had "easy holing." Journal, Jan. 3, 1839.

[14] As indicated on p. 21 of the company's 1836 "memoir," Alexander had expected the cost to be about $15 a ton. The average cost of a ton of pig iron produced in Great Britain in 1839 was $15.84. The average cost of a ton of pig iron produced in charcoal furnaces in Maryland in the same year was $23.92. In 1842 the Maryland and New York Iron and Coal Company produced pig iron in its coke-fueled furnace at Mount Savage for about $16 a ton. Alexander, *Manufacture of Iron*, pp. 170–171; and Maryland and New York Iron and Coal Company, *Statement of the London Committee*, p. 12.

The cost of wages per ton ($2.68) at Lonaconing in 1839 may be compared with $1.92 per ton for a furnace in Wales,

Estimated Raw Materials for 10 tons of Pig Iron per Day

40 tons of ore @ $3.00	$120.00
40 tons of coal @ .56¼	22.50
limestone	10.00
Cost of Materials per Day	$152.50
Cost of Wages per Day	26.80
Cost of 10 tons of Pig Iron	$179.30
Cost per Ton	$ 17.93

April 10.—Duvall's men gearing saws on the gang frame, in consequence of which the delivery of the mill is postponed. The following rules are also drawn up by Mr. Alexander and directed to be announced by the underground agent the next day: [*A*, *B*, *D*, and *F* levels would be closed. *C* level would be worked at $3 per ton of clean ore ready for roasting, with the men furnishing their own powder and candles. They would no longer be given laborers to dispose of the non-productive refuse which accumulated as they dug the ore. To the extent that places were available, men from the closed workings would be employed in *C* level or in the coal mine. Except for pushing forward the main heading of the *C* level, there would be no allowance for dead work—putting in props, laying and repairing tram rails, and cutting through stone and clay to get to the layers of ore.]

April 11.—Milholland finishes his portion of Savage road. Mr. Preuss and corps return from Dans Mountain, having completed the second survey from Lonaconing to Buck Lodge [a distance of 35½ miles]. Ground reported to be very rough in some places. Mr. Pauer applies for a less laborious situation.

April 12.—Water let into the race. Rain at night.

April 13.—Rain continues—freshet—race bank leaking. It is stopped after some little exertion by Milholland's men.

April 14.—Sabbath.—Mr. Forrest (Methodist) preaches in the miners' church.

April 15.—Messrs. Alexander and Tyson depart for Baltimore. Men not at work, and about 40 give in their names as wishing to settle. Dr. Wundsch takes umbrage at the deduction of Dr. Hunt's fees ($30) from his monthly quota, behaves in a most violent manner, and is dismissed by C.B.S. from the company's employment. Letter to Mr. Alexander, requesting him to send a physician.

April 16.—Dr. Wundsch going about and very foolishly persuading himself that after all his misbehavior he can be permitted to stay here.

April 17.—Duvall applies for money. Let him have $100 and refer him to the president of the company for the supply of all future wants. Letter to Mr. [Thomas] Perry [attorney] at Cumberland, praying writs against Benjamin Thomas, Roger Williams, and Thomas Phillips (*Tiberias* men).[15] Mr. Harrison brings the officers [con-

operating at the same time period and producing at the same rate (70 tons a week). Alexander, *Manufacture*, p. 141.

[15] The *Tiberias* men still owed the company considerable amounts of money, principally for their transportation from Wales: Roger Williams, $162.13; Thomas Phillips, $154.36; and Benjamin Thomas, $248.31. Clerk's Docket to Oct. Court, 1839, Originals Nos. 1, 2, and 3. Allegany County Courthouse, Cumberland. See also acct. of John Williams, Sr., Clerk's Docket to Apr. Court, 1839, Originals, No. 60.

stables] Porter and [William] Ravenscraft, who serve the writs. The men give their promise to be at work in the morning.

April 18.—Miners again at work. Number of men in the mines, 16; ditto in the coal, 14. Duvall going on well with sawmill, but requiring the chief use of it himself. In consequence, the bricklayers at the store and Welch's men are much delayed in their work. Store not going on for want of joists. Gruber and Grauss employed in building abutments to a bridge over the race below the dam. Letter from Mr. Alexander, directing that Duvall have money.

April 20.—Duvall applies for money; he receives $200. Dr. Wundsch still prowling about Lonaconing. The miners are allowed 20 cents per car of 12 [cubic] feet of coal from the No. 4 vein, and the powder is reduced to $4.50 per barrel.

Twelve cubic feet amounted roughly to 10 bushels. The Maryland Mining Company was paying its colliers one cent a bushel for coal dug and wheeled out to the pit mouth in wheelbarrows. A good digger in an established mine could deliver 100 to 150 bushels a day.[16]

One wonders what the price of the powder was before it was reduced. A 30-pound keg of powder cost $3.25 in Baltimore. William Alexander's commission amounted to 2½ per cent of the costs of the goods. Freight rates between Baltimore and Lonaconing ranged from $1.50 to $2.00 per hundred pounds. At the former figure, the freight was 45 cents a keg, making a total cost of $3.78 per keg delivered at the works. At the higher freight rate, cost of carriage was 60 cents a keg, making a total delivered cost of $3.93.[17]

April 21, Sunday.—Preaching by Messrs. Chambers and Venable. Letter at night from Mr. J.H.A. Mr. A. does not seem satisfied with the slow progress of the negotiation with Buskirk, and gives directions to commence a more decided course by the claim and assumption of possession of his coal drift, the timbering of the mines to be destroyed, and force, if attempted, repelled by force. Information that a new doctor is coming.[18]

April 22.—Letter to Mr. Richard Wilson, begging his interference to prevent the coming of Dr. Hermann, who from inquiry seems to be in every way unsuitable. All the Germans know him, and none of them speaking well of him. Furnace was fired on Saturday. [To be sure that no moisture was left in the masonry of the stack, the furnace was filled with coal or coke, and a fire was kept up until the founder was satisfied that the walls were

sufficiently dry.[19]] The coal appears to clinker, and Mr. Hopkins thinks that the furnace will require very active attendance to prevent it. Bridge house going on well. Letter at night from Mr. Alexander. He disapproves of the allowance to the miners for coal. Mr. Preuss has been complaining of the removal of Mr. Pauer from his party for other purposes. Mr. A. now directs that he resume his place in the party.

April 23.—Mr. [William] Shaw is sent for and instructed as to the most suitable proceedings in the case of Buskirk's coal opening. Letter at night from Mr. Alexander. Duvall has been drawing money both in Baltimore and here. Mr. A. directs that he shall have no more.

April 24.—The furnace has got heated, and cracks appear on the northeast and southwest sides. The t[r]unnel head is forced up at least 3 inches by the expansion of the stone of the inwalls. The discharged miners who linger about are sent off the premises, with exception of John Thomas, whose wife is ill.

April 26.—Mr. Graham has made an advance of $285 to Abraham Young [local farmer] to purchase 3 yoke of cattle which are eventually to be his property, and C.B.S. writes to Russell in Cumberland, ordering him a wagon. Young contracts to haul logs to the mill for 30 cents per hundred feet and to refund to the company one-third of his monthly earnings until his cattle and wagon are paid for. Final settlement for the Savage road! Total cost of [7½] miles, $[2,814], being $189 above the estimated sum of $350 per mile, the road thus costing per mile $[375.20]. The extra cost arises from improved construction in the turns, and from unforeseen difficulties in cutting.

April 27.—Preparations for blast continued. Tuyere pipes nearly ready for covering. Much activity in the coke yard—limestone breaking, mine burning, etc. C.B.S. much troubled with headache and unable to attend to business. Mr. Serpell attending to tuyere pipes and other matters of construction.

April 28, Sabbath.—Levi Harris preaches. C.B.S. still much indisposed.

April 29.—Orders given to Arner to make a filter box for the upper end of tuyere pipes. [This was a two-compartment filter bed 18 feet square and 8 feet high set on the bottom of the race just above the headgates. The upstream compartment contained a bed of sand and gravel 18 inches thick. After filtration the water flowed into the downstream compartment, which contained, and thus protected, the openings of the pipes.]

April 30.—C.B.S. and J. Hopkins engaged in measuring [ore] preparatory to settlement. Upon a comparison of the time, as kept by the corporals, with the quantities of ore gotten out, it appears that the men have realized $1.50 per day, and the boys half price. The miners have evidently not exerted themselves, and the fact now appears to be settled [that] the No. 4 ore can be got at $3 per ton. C.B.S. still much indisposed.

May 1.—C.B.S. confined to the house with indisposition. Tuyere pipes are laid, but the supply appears to be very small. Some evil-disposed persons the night previous had let the water from the race into the trench for the pipes and injured some of the joints. Conrad and Milholland's men engaged in repairing them. Cracks in furnace enlarge.

May 2.—C.B.S. still much indisposed with bilious symptoms and unfit for business. Welch now appears to have but 2 men, and is told in consequence that the contract for the superintendent's house is vacated. Mr. [Ignatius] Jarboe [a carpenter in Romney, Virginia] is

16 Katherine A. Harvey, *The Best-Dressed Miners* (Ithaca, 1969), p. 132.

17 Daybook May 8, 17, and 23, and June 16, 1838. Welch and Alexander Record Books, Md. Hist. Soc.

18 Henry Hermann, the new doctor, was born in 1812 near Hesse-Cassel. He was a graduate of the University of Marburg, and a lecturer there until he emigrated to the United States. George Bennett, "History of Bandon," *Oregon Hist. Quart.* 28: p. 334 (Dec. 1927). One supposes him to have been a fully qualified physician, since in 1839 he was admitted to membership in the Maryland medical society (Medical and Chirurgical Faculty of the State of Maryland). Eugene F. Cordell, *The Medical Annals of Maryland* (Baltimore, 1903), p. 436.

19 Overman, *Manufacture*, p. 164.

written to by C.B.S. and offered the contract for super-intendent's house.

May 3.—C.B.S. confined to bed. Mr. Serpell attending to business.

May 4.—C.B.S. again out. but not able to see distinctly. Mr. Jarboe arrives from Romney. C.B.S. much indisposed, but compelled to converse much with him touching his contract for the new house.

May 5.—Mr. Alexander arrives from Baltimore. C.B.S. much indisposed and still unfit for business.

May 6.—C.B.S. too much indisposed for any kind of business. Mr. Alexander is informed of his (C.B.S.'s) inability to keep the regular journal of Lonaconing transactions and undertakes to perform that duty himself.

Alexander takes over journal:

May 7.—In consequence of the sickness of C.B.S., J.H.A. continues the memoranda from this day. The furnace was found charged with cokes, and though cracked on all three external sides, yet not injured. Indeed, the heavy charge of cokes seems in itself a more injurious burden to the lining etc. than a charge of mixed iron and flux would have been. [Directions to take up the tuyere pipes and find out why the supply of water is insufficient.]

May 8.—The pipes for the water tuyeres are now found to deliver 14 gallons per minute. They are therefore directed to be covered in. Mr. Hopkins presents himself in bad spirits about the blast, and the matter is therefore taken in hand by the "gods" themselves, the effect of which interposition is that the northeast half of the bridge house is floored before night, and matters put on a going train for the balance and for the molding house also.

May 9.—At night (9 P.M.), amid much hurrahing and other demonstrations of lively interest, and by the light of certain candles fancifully arranged at the expense of the mental ingenuity and capital of the Messrs. Hopkins (but the combustion of which went on at the expense of the company), J.H.A., C.B.S., and R.G., in the order of their names, performed the important functions of putting in the first ferruginous charge of the furnace. From prudence or some other motive, no speeches were delivered, but a procession of human figures in a high state of excitement passed over the scenes to the ale house of D. Hopkins, where presently was commenced the imbibition of sundry quarterns of ale which, like the other combustible matter before mentioned, passed from the ownership of said David to the stomachs of the members of the late procession at the expense of the company. The curtains were drawn at 11 P.M.

May 13.—Mr. Jarboe remains to make the pig patterns. Molding house nearly finished.

May 14.—Anxious expectation of the blast being put on. [The blast was put on gently after the ore charge had sunk down to the hearth and little drops of molten iron could be seen in front of the tuyeres. A new furnace would not be ready to receive a full burden and a strong blast until it had been in operation for a week or two.[20]]

May 15.—Still looking for the blast. Molding sand all leveled off in the molding house.

May 16.—Today at 11¼ A.M. the blast is put on by J.H.A. in person at the northeast tuyere and by D. Hopkins at the southwest tuyere. C.B.S. very sick through the day. Furnace making excellent cinder [an indication of the quality of the iron.]

May 17.—C.B.S. unfortunately sick today. At 6½ A.M. the furnace makes its first run out of iron of good quality. This being the finishing stroke of a large part of the

FIG. 10. Furnace in Blast. Harper's *Monthly*, Jan. 1875.

subject of these entries, nothing more need be expected today except notes of admiration. ! ! ! ! ! ! ! ! ! ! ! ! ! ! ! ! !

In running out, the blast was shut off, and the plug of clay in the tap hole at the bottom of the hearth was pierced with a long iron bar. The molten metal flowed out into previously prepared pig molds in the sand floor in front of the furnace. When the run-out was finished, the clay plug was restored, the blast put on again, and furnace operations proceeded as before for twelve to twenty-four hours, when the furnace would be ready for tapping again. The furnace was constantly refilled while in blast so that the charging materials were always level with the top. The tap hole was at the right side of the tymp arch; the cinder ran out into cavities at the left side of the arch.[21]

May 18.—Yesterday, being occupied in congratulations etc., passed off very well. The furnace cast in the morning and at 5½ in the evening. Today the furnace again casts at 10 A.M. with improving iron.

May 19, Sunday.—The northeast water tuyere given out and taken out and replaced.

May 20.—Tuyeres so leaky that a cold blast is substituted for the hot one. Also the arch of the southwest hot air furnace falls in. [Abraham?] Loughridge [pattern-maker] finishes the pattern tonight for the tuyeres, which will be made of cast iron.

Shaw resumes journal:

May 21.—C.B.S. again able to pay some attention to business. Furnace going on pretty well, but the delays in

[20] *Ibid.*, pp. 165, 186.

[21] Tomlinson, **2**: pp. 81–82; and Overman, *Metallurgy*, pp. 514–515.

FIG. 11. Filling a blast furnace. Harper's *Weekly*, Nov. 1, 1873.

changing tuyeres and the change from hot to cold blast has changed the character of the iron from "foundry" to "forge-pigs." [No. 1 foundry iron was best for fine castings. The forge-pigs (No. 3) could be used either for forge or foundry and were used for making heavy castings.[22]] J. Hopkins and D. Hopkins, as well as Mr. Schmidt, directed to furnish lists of their men with a view to diminish the expenses of the works, which by gradual additions to their forces have, during the protracted illness of C.B.S., become enormous.

May 22.—Furnace making "forge-pigs" about 6 tons daily. Mr. Duvall returns from Baltimore. Preparations for setting boiler and making chest to steam shingle block.

May 23.—New tuyere cast. From appearances it will do very well.

May 24.—Mr. Schmidt's corps is reduced to 17 men, viz. 2 wagoners, 2 ox drivers, 3 carters, 3 laborers for sawmill, and 7 miscellaneous. August Weisskettle dissatisfied, and his men complaining because they are not to have work all summer. Weisskettle is directed to finish the store.

May 25.—Mr. A. departs at 1 P.M. for Baltimore. The citizens of Cumberland and Frostburg much elated with the success of the iron works on George's Creek.

May 26. Sabbath.—Welsh and Methodist preaching.

May 27.—Furnace still making No. 3 iron, about 4 or 5 tons at a cast, but much embarrassed by the burning out of the tuyeres. D. Hopkins gives in the following list of workmen necessary for the operations of the furnace: 1 founder, $110 per month; 3 keepers @ $60, $180; 2 fillers @ $50, $100; 2 cinder fillers @ $40, $80; 2 helper fillers @ $40, $80; 2 limestone breakers @ $37.50, $75; 1 coker, $45; 2 coke drawers @ $30, $60; 2 barrow fillers @ $37.50, $75; 2 helper boys @ $18.75, $37.50; 4 coal setters @ $37.50, $150; 3 bridge stockers @ $37.50, $112.50; 2 molders @ $50, $100; 1 mine burner, $40; 3 laborers @ $30, $90; 2 engine men @ average $55, $110; 2 firemen @ $37.50, $75; 2 stove tenders @ $37.50, $75; total monthly wages, $1,594.50. [This is almost twice what J.H.A. estimated on April 9. Shaw next lists wages paid at Varteg furnace, near Pontypool, Wales, quoting Dufrénoy's *Voyage Métallurgique*[23] 1: p. 349. He concludes that the monthly

wages at Lonaconing are "more than double the prices paid at the Varteg furnace, the difference being $931."]

May 28.—Two new tuyeres are cast this morning, and 5 tons of dark grey (No. 3) iron. Directions given to make 3 trams for limestone, one iron tram for furnace cinder, to make irons for 5 pumps, and to prepare one of the tram wheels for a pattern to cast from. Rees Jones proposes for plastering three coats 15 cents per yard, or 12½ and the company to find the laborers, and 15 cents for rough casting—very modest! He only values his labor at the cost of labor and materials in New York!

May 29.—Both tuyeres have burnt out during the night, and we must now blow "cold" until more water can be got to the tuyeres.

May 31.—The men at the furnace want more wages. Visit from the great horseleech [D. Hopkins?] on that subject.[24]

June 1.—D. Hopkins has presented and been paid $78 for beer drank at the blowing in. While making arrangements for supplying the tuyeres with water, it would not be amiss to lay a line of pipes from D. Hopkins' house to supply the 40 throats at the furnace with beer.

June 2. Sunday.—D. Jones, the independent preacher, preaches in the afternoon in English and Welsh.

June 3.—Loughridge making patterns for castings.[25] New arrangements are made for getting rid of the cinder at the furnace, the accumulation of which has become very inconvenient. Morning cast at 10 A.M. = 5½ tons of best foundry iron. Under this high pressure at the furnace, the stock of ore must decline very fast. Visit from Professor Johnson[26] of the Franklin Institute and Mr. Oliphant of the Fair Chance works.

June 4.—Letter to Mr. Alexander, explaining the heavy demands made in every quarter upon the finances of the company. Directions to Mr. Schmidt to employ three men at night in sending down limestone by the incline. Schmidt's men contrive to send down one load of limestone at night and to break two trams. Blacksmiths employed in making new trams for the limestone.

June 6.—Teams employed in hauling lime, sand, brick, cinder, and logs to sawmill, and a cart and horse in clearing the molding house of the stock of pig metal. J. Hopkins has returned the work for May, whence it appears that 300 tons of No. 4 ore and 600 tons of No. 4 coal have come out during the month from the C level. Total cost of the ore per ton, including cleaning, hauling, and stacking up, $4. Ditto of No. 4 coal, 33⅓ cents. The consumption of coal at the furnace seems to have been about 7 tons to 1 ton of metal. At the Calder No. 3 furnace[27] the consumption under similar circumstances (coke and cold blast) was 8½ to 1, while with hot blast and raw coal the consumption was less than 3 to 1. The advantage of the latter method is not only in the smaller consumption of materials, but in the reduced number of hands, as the entire list of cokers, coke drawers, and coke setters—6 or 8 men—will be at once dispensed with. Mr. Oliphant's furnace at Fair Chance runs out about 4 tons daily, and the men employed there, according to his statement, are in number 13, as follows: 2 keepers, 1 mine hauler and

---

[22] Tomlinson, 2: p. 82.

[23] Ours-Pierre Dufrénoy, *Voyage Métallurgique en Angleterre* (2 v., Paris, 1837).

[24] Horseleech: one who makes inordinate and endless demands.

[25] Shaw also ordered patterns from Winans. W. Alexander to R. Graham, June 13, 1839. Letter Book 2, Welch and Alexander Record Books.

[26] Walter Rogers Johnson (1794–1852), authority on coal and iron. His visit to Lonaconing is described on p. 8 of his *Notes on the Use of Anthracite in the Manufacture of Iron* (Boston, 1841).

[27] This furnace was near Glasgow. The figures are from Dufrénoy, 1: p. 405.

clamp filler, 1 mine burner, 1 limestone breaker and bridge stocker, 1 cinder hauler, 6 colliers [charcoal makers] employed in the coal ground.

June 7.—Clark, the smith and engineman, complains of the hardness of his duty and wishes that the fireman and coal hauler may divide their duties so as to lighten his, which from the joint cares of the engine and smith's shop are no doubt burthensome. Mr. Alexander directs the burthen to be diminished at the furnace, and that the making of tram plates and railway iron shall commence as soon as possible.

June 8.—Letter from Mr. Alexander at night giving C.B.S. full discretion with regard to the general management of the works, only restricting disbursements to $10,000 per month.

June 9, Sabbath.—D. Jones preaching in the miners' church.

June 10.—Conversation with D. Hopkins on the subject of reduced expenditures at the furnace. He promises to give all needful aid in effecting that object. [He] is directed to bring the yield of the furnace to 6 tons per day and the quality to dark grey or mottled suitable for railway iron. Lewis, Vivian, and Rice (Cornish miners) are threatened with violence by some of the Welsh miners, and wish on that account to leave the place.

June 11.—Duvall is engaged in trials of [sawmill] machinery—cross cut and ripping saws and lathing machine—and in fitting up the steam chest for the shingle blocks. Wortman building the boiler stack.

June 12.—Visit from Mr. [Charles B.] Fisk, engineer to the Chesapeake and Ohio canal. He wishes to see Mr. Alexander and to make arrangements for the union [of] the interests of this company and the one he serves in regard to the depot at Buck Lodge. The crack in the furnace enlarging, and some of the nuts of the holdfasts drawing the threads of the screws having given way, Loughridge is directed to make patterns for new plates (to be of less thickness, thereby enabling the nuts to take hold of new threads.)

June 13.—Visits from General Duff Green, Mr. George Greene, Caspar Wever, Esq. [superintendent of construction for the Baltimore & Ohio Railroad], and Dr. Smith of Cumberland. Duff Green applies for specimens of ore and coal, which are sold to him for $13. He forks up the money—*mirabile dictu!* Good omen of success when gold is extracted from so barren a source.

June 14.—Pagenhart receives a severe injury by a fall from the gallows frame of the coal incline in refitting the drum. Dr. Wundsch is sent for at the instance of the Germans, who do not seem to have faith in Dr. Hunt's skill. C.B.S. endeavors, but too late, to prevent his being sent for. Dr. Wundsch arrives in the course of the night. Dr. Hunt had previously bled Pagenhart.

June 15.—Pagenhart is better, but seems to have sustained some severe internal injury. Hill, the bailiff, is sent round with a circular to the workmen, informing them that no restriction will be put upon their right to send for a physician other than the one employed by the company and paid from the medical fund, except that physician be Dr. Wundsch or some person of notoriously immoral character, in which case the right can only be exercised at the expense of the individual's employment by the company. Cast at the furnace in the morning—4 beds of pigs = 4 tons, and a large plate weighing probably 1 ton for the purpose of pressing out cinders from the runnel. Mr. Hopkins much discontented with the condition of the temporary molding house, which leaks badly and wets the sand. None of the castings have, however, blown, the sand being extremely clean and retaining very little moisture.

Fig. 12. Hauling cinder away from blast furnace. *Harper's Weekly,* Nov. 1, 1873.

June 16. Sunday.—Messrs. Alexander and Wilson arrive on foot from Frostburg. English service by D. Jones, independent preacher. None of any other persuasion now at Lonaconing.

June 17.—A riot has occurred at D. Hopkins' grog shop, and one of the miners, William Williams, shews himself much beaten and disfigured and charges the family (males and females) of Hopkins with having made a simultaneous assault upon him. John Hopkins claims the discharge of William Williams and is told that the matter has been referred to Mr. Alexander. D. Hopkins is also referred to Mr. A. Both Hopkins much enraged and threatening to leave the place unless Williams is discharged. John Hopkins is directed to remove Richard Hughes from his situation as corporal and put him in the coal or the mines as might be convenient.

June 18.—Mr. Alexander takes D. Hopkins to task about his grog transactions. He pleads guilty to the charge of selling "peppermint," and promises not to offend in like kind hereafter. Mr. A. thereupon decrees the discharge of William Williams for disorderly behavior in Hopkins' house. John Hopkins is sent for by C.B.S., and inquiry made why the order for the reduction of R. Hughes to the place of an ordinary workman had not been made, said Hughes seeming to be going about as before with his badge of office (his stick). J.H. explains the occurrence about as well as he usually does his transactions—that is, that he had designed, as usual, to postpone what he had been directed to do. He is told to remove Hughes from his post of corporal at once. He then requests that his own account may be settled at the same time, proposing then to leave the company's service. C.B.S. expresses his perfect satisfaction with the wish of J.H. It will not be difficult to find an agent of more skill and less conceit.

June 20.—John Hopkins brings in his map of the iron and coal mines, and communicates his thoughts in regard to the proper method of working them—in compassion, no doubt, to the ignorance of his auditors, J.H.A. and C.B.S., who of course can know nothing of the proper method of ventilating mines.

June 21.—Mr. [David] Hopkins represents the furnace as now ready to make tram rails. Mr. Alexander engaged in investigating the most advantageous form for that purpose in the evening.

June 22.—Messrs. Alexander, Wilson, and Shaw visit the coal level and limestone quarry. Find the latter nearly

worked out. Dr. Hunt brings information that Dr. Hermann has arrived at Frostburg which does not, however, appear to be the case, as Rupprecht's carriage brings but two passengers. Rupert's two passengers are the Reverend Mr. Owen and lady, who are quartered for the night at J. Hill's, and notice is given that Mr. Owen will preach on the morrow.[28]

June 23, Sunday.—Morning and evening service by Mr. Owen. Dr. Hermann does not make his appearance.

June 24.—Mr. Wilson and Mr. Alexander depart. Morning cast—5 beds of mottled iron (about 5 tons) and about a ton of iron plates for bridge house floor. Mr. Duvall's accounts are examined by C.B.S. and adjudicated in conformity to written instructions from Mr. Alexander. The items of "extra" framing, hewing, and forgotten details of Mr. Duvall's first estimate [$5,000] (on which his contract was based) reserved for future adjustment. It appears that excluding the cost of material neglected in the first estimate and amounting to $716, the sawmill and machinery for lath and shingles have cost $9,348, or $10,065 if the materials be included. The whole cost to Mr. Duvall he represents to have been as follows: labor, as per workmen's receipts, $5,125; transportation of hands and self, $600; materials and patterns, $3,638; total, $9,363. His profits have been diminished, no doubt, by keeping 5 or 6 hands during the winter with very little to do. Mr. Duvall has been paid at various times $7,660. [Mr. Graham is instructed to give $900 more, making total payment to Duvall $8,560. Duvall claims $1,505 still due.]

June 25.—Mr. Duvall leaves Lonaconing. He leaves in the mill about 2 cwt. of screw bolts and an iron shaft intended for a lathe, C.B.S. promising to consider the propriety of taking them off his hands.

June 26.—John Hopkins seems not to understand that he had discharged himself [from] the company's services, and wishes to discuss the plan of future works in the mines. On being questioned, however, avows his having proclaimed his determination to leave. He is told to commence measuring up for June on Tuesday, and that Mr. Baxter, who has been engaged to take his place as underground agent, will attend him in the measurement. Strebig employed at the shingle machine, getting out shingles for the store. Welch is engaged in shingling store and getting out floors, fitting architraves to doors, and other joinery about the store. Mr. and Mrs. Owen leave Lonaconing.

June 27.—Masons at molding house have made a mistake in the arch of the great door and have to take it down. A committee of gentlemen—stockholders in the Chesapeake and Ohio Canal Company—visit the works. The coal level and the mines are visited by them, and much satisfaction expressed with the progress of the works. Furnace making forge pigs. Several Scotch and 2 German miners applying for work. They are told that there is no wish to increase the number of hands at this time, but that on the 1st July a new management will commence in the mines and that some of them may be employed if they choose to wait a few days.

[28] The Reverend John Owen, rector of Trinity Church, Washington, D.C. The English-born Owen, a former Congregational minister, was ordained as deacon on July 29, 1836, and as priest on July 31, 1837, by Bishop William M. Stone of the Maryland diocese of the Protestant Episcopal Church. *Journal of a Convention of the Protestant Episcopal Church of Maryland, 1836* (Baltimore, 1836), p. 10. *Ibid.,* 1837 (Baltimore, 1837), pp. 11, 13. See also Ethan Allen, *Clergy in Maryland of the Protestant Episcopal Church since the Independence of 1783* (Baltimore, 1860), p. 52.

June 28.—The furnace, through the neglect of the keeper (William Hopkins), has gotten clinkered during the night and is with much difficulty relieved from this dilemma by D. Hopkins. W. Hopkins is sent for to give an account of this circumstance, but does not make his appearance, being probably engaged in sleeping off his debauch of the previous night at Buskirk's, where it is supposed he was when the furnace scaffolded [became obstructed by partly fused materials above the tuyeres]. Jim Crow and other buffoons arrive at Lonaconing, and Mr. Hopkins requests permission for them to exhibit at his house, which is accorded on condition that he will be responsible for the good order of the assemblage. Sundry misdemeanors on the part of females, Welsh and German, in which no clear understanding can be obtained of the facts. Myer, the wagoner, is admonished that he must tame his shrew and keep better order in his house. Mr. Hill reports that A. Funcke is said to keep whisky in his house. He makes search, but without success, though it afterwards is fully proven to him by the testimony of Ledley's men that Funcke had sold liquor. Ledley has tried two experiments in the making of firebrick, the first with $\frac{1}{2}$ of pounded brick and $\frac{1}{2}$ clay, and the second with $\frac{3}{4}$ fireclay and $\frac{1}{4}$ clay from the brickyard. He fails in the former experiment, the brick melting completely in the furnace. The second experiment yields brick which takes a fine glaze and will probably stand fire well.

June 29.—The mason employed by Milholland at the molding house has again committed an error in building the arch to the great door. [The mason's error was allowed to stand.] Weisskettle's new hearth is finished, and the third course is laid together [without mortar] to ascertain its correctness. Weisskettle takes the plastering of the interior of the store at $12\frac{1}{2}$ cents, the company to find materials. Mr. Schmidt has ascertained that any quantity of hair may be had at Frostburg at 30 cents per bushel. Rees Jones (too late for his own wishes, but to the great satisfaction of C.B.S.) offers to take the plastering at $12\frac{1}{2}$ cents. He is told that Weisskettle had taken it 2 hours before.

June 30, Sabbath.—Preaching by David Jones.

JULY 1–DECEMBER 31, 1839

Surviving the financial panic of 1837, the United States enjoyed a period of recovery lasting from the spring of 1838 to the summer of 1839. Then the nation plunged into another credit crisis and three years of depression. One of the principal manifestations of the new crisis was the curtailment of English investments in American enterprises.

The president and directors of the George's Creek Coal and Iron Company, issuing a report on August 1, 1839, acknowledged the existence of "a fresh crisis . . . in monied affairs," but reassured their stockholders that the company still had funds for carrying on its enterprise "with vigor but little diminished."[1] The deposits of coal and iron at Lonaconing had exceeded all expectations; the blast furnace was producing pig iron of excellent quality; and the company was taking steps to move into more advanced stages of iron manufacture. A small cupola furnace already under construction would make fine castings from the stock of pigs on hand (about six hundred tons). The next improvement, expected

[1] GCC&I Co. Report 1839, p. 13.

to be accomplished within the year, would be the erection of a forge and rolling mill for the production of bar iron. The coal trade, even as early as this viewed as potentially the most important part of the company's operations, was unfortunately "kept in abeyance" by delays on the Chesapeake and Ohio canal. Therefore the contract for the railroad to the shipping depot on the Potomac would not be let until the spring of 1840.

In view of succeeding events, the report seems unduly optimistic. The blast furnace was blown out on August 17, and the force of miners and furnace men was greatly reduced. Throughout the remainder of 1839 wages were cut drastically, and operations were gradually drawn in, until at the end of the year the works were virtually closed. During most of this period a major part of the activity at Lonaconing involved the construction of three expensive buildings with little or no immediate income potential: the new store, which would have few customers now that most of the workmen had left; the molding house, intended for casting from the big furnace now out of blast; and the elaborate house for the superintendent.

This latter situation was not of the superintendent's making, since plans for the layout of the furnace complex and the village had been made by Alexander. However, largely because of Shaw's ineptness in handling people, the work progressed at snail's pace. The contractors ran up the costs, squabbled with each other about sharing new carpenters sent from Baltimore, and quarreled with their own journeymen.

From a distance it appeared to Alexander that everything else was also going wrong. The coal and iron mines were mismanaged; the cost of the new castings was too great; and to cap it all, the railroad surveys were inaccurate. For all of this, Alexander uncharitably blamed his superintendent.

July 1.—Masons at molding house had built up the pier belonging to the west door of the molding house upon an imperfect foundation and [had] commenced to turn the arch of the door when cracks were discovered. They are directed to take the whole down, and as the entire pier could not be founded upon the limestone below the furnace hearth on account of a wooden water pipe which carried off the tuyere water, a stone sill is directed to be made, and a plate of iron 3 feet broad and $2\frac{1}{2}$ inches thick is laid on the sill and on the projection of the furnace hearth so as to span the opening for the water pipes. Limestone is again scarce, and Mr. D. Hopkins is compelled to use the "bastard limestone" from the bottom of the quarry. Mr. George Greene, engineer, applies for leave to copy part of Mr. Preuss' map of the railroad surveys, and is permitted by C.B.S. to do so, but upon its appearing that same is for the use of Duff Green, a caveat is entered by C.B.S. until Mr. Alexander shall himself have signified his consent to such use of it. Mr. Greene leaves his map at Lonaconing and promises to see Mr. Alexander on his way through Cumberland.[2]

[2] George Sears Greene, a graduate of the U.S. Military Academy, resigned his commission in 1836 and become a civilian

July 2.—This day and yesterday are much occupied with new arrangements for the underground works. Four German, 2 Scotch, and 2 new Welsh miners are entered, and some changes are made also at the furnace. Mr. Baxter, the new "agent," presents his testimonials, whence it appears that he is at least a trustworthy and sober man. Daniel James applies for permission to sell beer. It is granted on condition that he sells no spirits and keeps good order in his house.

July 3.—Consultations with Mr. Baxter as to the best mode of reducing the price of the coal. This is difficult to accomplish except at the risk of blowing out the furnace [if a strike should cut off the supply of coal]. Mr. and Mrs. Alexander arrive, dine at Mr. Graham's, and proceed to W. Shaw's. Arrangements are made with some difficulty to get the proper supply of coal to the furnace on the 4th, the men wishing to keep holiday.

July 4.—Anniversary of American independence. Too much independence among the workmen to celebrate the day in any other way than by getting drunk, which many of them do, and for which any other occasion would have done as well. Mr. Alexander proposes for the morrow the commencement of a reconnaisance of the last railroad line to Buck Lodge, a duty from which C.B.S. thinks he ought to have been exempted on account of a well-known infirmity, but of which he determines to make no complaint until after having performed the duty.

July 5.—C.B.S. and Mr. Preuss set out for W. Shaw's, are met there by Mr. Alexander, and reconnoiter the line to Westernport. Sleep at Westernport, and hear there a disagreeable rumor of an accident at Lonaconing from the fall of one of the arches at the new molding house.

July 6.—Reconnaisance continued. Ground very difficult for 3 miles. Balance of the line to the neighborhood of Paddy Town [now Keyser, W.Va.] good. Sleep at Wesley Dawson's near Fort Hill.

July 7, Sunday.—Mr. Preuss returns to Lonaconing. Messrs. Alexander and Shaw remain at Dawson's.

July 8.—Mr. Preuss returns to Dawson's, and the reconnaisance to Buck Lodge is completed this day. The line of location will be generally upon good ground with the exception of 4 difficult miles near Westernport and a few miles upon other parts of the line in approaching ravines. Some changes must be made upon the latter part of the line, but a sufficient evidence is afforded in the line already run that a cheap road may be constructed from Lonaconing to the depot on Potomac. C.B.S. much fatigued by this trip.

July 10.—J.H.A. and C.B.S. return to Lonaconing. The retaining wall next to the furnace stack at the south end of the molding house had fallen and several men been hurt, one dangerously. The furnace had also got to bad work from a deficient supply of flux and coal, and was found to be making "bright and white iron" [not suitable for fine casting]. Thomas Powell and John Williams, during the absence of C.B.S., have been brutally ill-treating their wives. The wife of George Miller has also been detected in a disgraceful situation by her

engineer. Much of his work between 1836 and 1856 concerned mining. Amer. Soc. Civil Engineers, *A Biographical Dictionary of American Civil Engineers* (New York, 1972), p. 52. Greene's 1842 map of the Frostburg coal field, showing iron ore openings and the Mount Savage works, regrettably does not cover the George's Creek area southwest of Frostburg. Eleutherian Mills Historical Library, MS Accession 118.

husband, from whom she also receives a beating. No chivalry in Lonaconing.

July 11.—Furnace making mottled iron [suitable for railroad iron]. Loughridge making patterns for tram plate flasks [the wooden or metal frames holding the sand for the molds in casting] and core boxes for tram wheels. Milholland is directed to remove the ruins of the fallen arch at the molding house.

July 12.—Many miners applying for work from Pottsville. Mr. Baxter is directed to endeavor to find 12 or 15 colliers who will consent to work the coal at 50 cents per ton, [so] that an attempt may be made to put the rest at that price. Furnace much embarrassed in its working by the bad quality of the flux.

July 13.—Messrs. J.H.A. and C.B.S. visit limestone quarry. Directions to D. Jones and G. Williams to open another pit, which it is hoped will find limestone of good quality nearer to the surface, the present head of stripping [overburden] being about 9 or 10 feet.

July 15.—This day has been appointed to examine into the reported cases of brutality to wives. George Miller and Thomas Powell are accordingly dismissed the company's service, and John Williams reprimanded and fined $10.

July 16.—"Foundry iron" again, and the door casing to one [of] the new [store] vault doors cast at night.

July 17.—Furnace still making good iron, but Mr. Hopkins much troubled by the misconduct of John Lewis and David Thomas [keepers], who are accordingly fined, the former $10 and the latter $3, for their neglect. The flasks for tram wheels and plates are cast today.

July 18.—Water getting very low, and much difficulty found to keep a sufficient head in the pond to cover the origin of the tuyere pipes without shutting the headgates and stopping the sawmill, the dam leaking badly at the foundation. [Michael Milholland contracted to repair the dam for $200, paid 30 days after the work was finished, provided that the dam and abutments had been watertight for that length of time. The company would furnish cement and lime.]

July 19.—Young, the lumber-getter, is taken to task for overcharging the company for the use of his oxen. Sundry affairs of police: Ant. Funcke, Edward Lewis, George Stark, and William Treasure fined $5 each for violation of the hog ordinance. Many applications for work from colliers and miners, the works at Pottsville being very dull. Mr. Baxter is directed to make the best possible use of this opportunity to reduce the wages of the colliers and to take new miners into the No. 12 measures if they will contract at $3.50 per ton for ore.

Pottsville, Pennsylvania, a center for anthracite coal mining, also supported a number of furnaces and forges. The panic of 1837 had caused the suspension of many collieries, and even in 1839 business had not recovered. Many mines were idle or working part time, and miners were leaving the region to look for work elsewhere.[3]

July 20.—Messrs. Alexander and Shaw making out estimate for the Lonaconing railroad. The result in round numbers is for construction of 42 miles, $300,000, and $20,000 for depots and water stations, making an average cost of $8,000 per mile.

July 21, Sabbath.—Preaching in the morning by Mr. Jones, in the evening by Mr. Skinner.[4]

July 22.—Ledley hands in his brickmaking account. He has made this summer 440,000 bricks, which at $4.25 [a thousand] amounts to $1,870, and 1,500 firebricks at $20 = $30. Total cost of brick for the season, $1,900. The men at the furnace who had been denied by Mr. A. the allowance of beer for casting bring forward a new discontent on the subject of the fines imposed for neglect, and seem rather mutinous. Mr. Alexander, preparing to leave for Baltimore, leaves a memorandum for Mr. Matthews concerning the resurvey [of Commonwealth].

July 23.—Mr. Alexander and lady leave Lonaconing.

July 25.—C.B.S. determines to retain no more colliers at the present extravagant rates than will be sufficient to supply the furnace, and for that purpose consults the underground agent and [the] founder. The present number of colliers is 28, of whom 11 can very conveniently be dispensed with. Mr. Baxter is directed to bring his time book in the morning [so] that selections may be made of the men to be discharged.

July 26.—Eleven colliers are selected from the head of the list to be discharged. Bad news by last night's mail from England in regard to the money market.[5] Mr. Wilson in a letter to Mr. Graham seems to be of opinion that we must "blow out." This measure would not be inopportune on other counts. The price in the coal might be better regulated, new work opened at leisure in both coal and iron mines, firebricks made for new inwalls to the furnace; and by the operation of the small cupola which it has been some time in contemplation to build, all the tips and underground works may be furnished with iron tram rails. Evening cast at 6 P.M.—4 tons of No. 1 pigs and one of the plates of the little cupola.[6] Very hard tapping in consequence of the furnace having slightly cooled during the operation of cleaning boilers and adjusting steam engine, which took place yesterday. Loughridge drunk! must be discharged.

July 27.—But little going on today in consequence of some equestrian performance exhibiting at Frostburg, whither most of the Lonaconing population has gone. Loughridge still too much obfuscated to be properly sensible of reproof. Jarboe has framed the entire lower story of the new house. In the afternoon he applies for $50, which is given him with the information that no further advances will be made until he has brought

---

[3] Swank, pp. 195, 355–359; and Pottsville *Miners' Journal*, Aug. 24, 1839.

[4] Denomination not known. George Greene's 1842 map shows "Rev. Mr. Skinner" about ¾ mile northeast of Frostburg. The only church in the vicinity was the Lutheran meeting house.

[5] The *Great Western* arrived in New York on July 22, bringing news that the London money market was in a "state of pressure and disquietude," that the bullion supply of the Bank of England was greatly reduced, and that American securities were "unsaleable at any price." The *British Queen*, arriving on July 28, reported continuing pressure in the English money market. Baltimore *Sun*, July 24, 25, and 30, 1839.

[6] A cupola furnace, much smaller than the tall blast furnace, resmelted pig iron and scrap iron for casting. Up to this time the George's Creek company had been casting directly from the big furnace in addition to running out beds of pigs. The small furnace would use up the stock of pigs and would require fewer men to operate it. One type of cupola furnace consisted of a cast-iron cylinder 5 to 8 feet high and 30 to 40 inches in diameter set on a brick or masonry base 20 to 30 inches high. It was open at the top to allow the escape of flame and gases and to permit charging with metal, coke, and limestone in proper proportion. Tomlinson, 1: pp. 344–345.

his work beyond the payments. Jarboe is also told that he must stop his store accounts.

July 28, Sabbath.—Welsh preaching in the morning. No afternoon service.

July 29.—The "doomed" colliers allowed to square up their work, which will probably be done in time to allow of settlement on Wednesday. Castings for cupola going on; iron first rate. C.B.S. compelled to go to bed at noon by bilious affection.

July 30.—C.B.S. confined to bed by severe headache. Much indisposition among the people of the place, chiefly cholera and bilious affections. Dr. Hermann reports 32 sick.

July 31.—W. B. Rogers, geologist of Virginia,[7] and his assistant, Mr. Hayden, visit the works and dine with Mr. Graham. Loughridge, who has been admonished and told that he will only be continued in his place until a successor can be appointed, is today sober.

August 1.—Settlement with the discharged colliers. Letter at night from Mr. Alexander. C.B.S. again bilious and taking blue pills [pills of prepared mercury, probably calomel, used as a laxative].

August 2.—Mr. Baxter complains that the colliers do not begin their work until late in the morning, and that much time is in consequence lost in the coke yard, no coal coming down the incline until after 9 A.M.

August 3.—Letter to Mr. Alexander with a proposal to "blow out" the furnace and place the establishment upon a more economical scale by which the monthly expense may not exceed $6,000. The little cupola with 7 hands to substitute for the high furnace [with 38 hands] in casting tram rails. Firebrick to be made for relining the furnace and building coke ovens and mine kilns, and a good stock of coal, ore, and flux to be collected before resuming blast, with a view to be independent of a strike among the colliers or miners. An order is published to compel the colliers to observe the same hours as the furnace hands, beginning at 6 A.M. The foundation of the little cupola is begun opposite the center of the southwest arch of the molding house. Loughridge and Wagner making patterns for the blast pipes to cupola.

August 4, Sabbath.—D. Jones preaches in Welsh and English. Unusually quiet at Lonaconing, though a "ring fight" appears to have taken place at Buskirk's between some of the furnace men and the colliers.

August 5.—The colliers appear in a body and refuse to comply with the order of the third in regard to time, demanding 7 A.M. as the hour for beginning. After due reflection, it is determined to blow out the furnace and dismiss all the colliers except the *Tiberias* men. Mr. Jarboe has 2 more hands and is going on faster. Welch complains that Jarboe has seduced away his hands. Letter from Mr. Alexander, requesting to have a list of the furnace hands, colliers, miners, and other workmen employed by the company.

August 6.—C.B.S. indisposed. Mr. Pauer employed in making out the July list of the company's hands.

August 8.—Mr. Alexander is requested to send Welch 4 journeymen, and to send up a painter for the new store. Foundation of cupola still being dug—present depth ten feet below hearth. Ground appears hard two feet lower. Part of the blast pipes for the cupola cast at 9 P.M.

August 9.—Letter from Mr. Fisk [chief engineer] of the Chesapeake and Ohio canal ordering 150 cast-iron

dowels for lock walls.[8] Furnace making good iron (quality No. 3). Mr. Baxter detects B. Thomas and P. Parry in attempting to defraud the company in various ways, and other miners in the same attempt in the measurement of the No. 4 coal. They are docked in the gross sum of $70 in the July settlement, and B. Thomas and company $43 for false measurement of ore.

August 11, Sabbath.—No preaching except Welsh.

August 12.—B. Thomas drunk today and complaining of ill will towards him on the part of Mr. Baxter.

August 13.—C.B.S. discovers that he must visit Baltimore, and makes preparation to do so. Directions given to Mr. Hopkins to blow out the high furnace on Saturday, the 17th.

August 14 to 21.—C.B.S. leaves for Baltimore and is absent one week, in which interval the furnace has been blown out. The cupola is finished. The furnace hands except 8 are paid off and retire, most of them without disorders on the place, though a "ring battle" is said to have occurred at Buskirk's on Sunday, 18th. Messrs. Matthews and [Hanson B.] Pigman are engaged on the ejectment surveys [determining boundary lines of Buskirk's property adjoining Commonwealth], and from Mr. Matthews' report the case is in rather a "blue" way, Mr. Buskirk having proved his corner, or beginning.

August 22.—No success among the Welsh in letting the coal at 50 cents per ton. Four Germans have, however, taken the work at that price, and C. Blatter, Jr., is only waiting for a partner to do the same.

August 23.—Clark [blacksmith] is sent for and admonished for his connection with the recent disorder at Buskirk's, and directions are given to him to send away 2 smiths and 2 helpers. Mr. Schmidt is directed to keep the time of the blacksmiths and carpenters and to discharge Pagenhart and Reuter at the end of the month. Mr. Thomas Alexander arrives from Baltimore at night. Mr. Graham receives a letter from Mr. J.H.A., and C.B.S. none.

August 24.—Men at furnace fitting up blast pipes to cupola. Mr. D. Hopkins sick. Two carpenters arrive from Baltimore. Riot among Germans at night.

August 25, Sabbath.—Very quiet this morning. A notice is posted requesting the company's officers to forbear discussing the conduct of J. Hill, the bailiff, as subversive of discipline, Hill having rather unnecessarily rang the bell on Saturday night. Mr. T. Alexander does not seem to approve the notice, and advises its removal. C.B.S. requires previously, however, to be convinced of its impropriety, and in default of such conviction is compelled with all proper deference to Mr. A's judgment to let it remain.

August 26.—Mr. Jarboe is informed that he cannot have [the new carpenters] without some arrangement with Mr. Welch, which he effects upon promising Mr. Welch the next two (who are said to be on the way), and that he will also spare to Mr. Welch two of his men to aid him in framing the bridge house roof.

August 27.—Health of the place not good. C.B.S. again slightly indisposed. C. Fazenbaker is sent to to continue the church, but keeps out of the way. Preparations continue to blow in little cupola.

August 29.—Mr. Baxter announces that all the Scotch miners from Frostburg will go into the No. 12 measures at the prices lately announced—$3.33 per ton. The same notice proclaims that after Tuesday the 10th September, the large vein coal will be put at 50 cents the ton of 28 bushels, and that those indebted to the com-

[7] William Barton Rogers (1804–1882), professor of natural philosophy at the University of Virginia. He and his brother, Henry D. Rogers, were authorities on coal formations. *Dict. Amer. Biog.*

[8] The Fisk correspondence in the Chesapeake and Ohio Canal papers at the National Archives does not include this order. There are no copies of letters sent after July 3, 1839.

pany and refusing to take the price must then be prepared to settle. This long warning was advised by Mr. T. Alexander.

August 30.—Surveys of Commonwealth close. The impression of Messrs. Matthews and Alexander seems to be that Buskirk has proved possession upon a part of the disputed lots, but will also lose a part and have therefore to defray the costs of suit. Mr. T. Alexander and C.B.S. visit various parts of the work and confer upon measures of policy. Letter at night from the president, Mr. J.H.A., who writes from New York on the point of departure for England.

Having succeeded in selling a large amount of the company's common stock in England in the spring of 1838, Alexander was now bound for London to try to market £50,000 of the company's bonds, secured by a deed of trust on Beatty's Plains granted to Charles Christmas and Rufus Prime of New York, and Richard Wilson of Baltimore.[9] Alexander was not able to sell the bonds, but left them in the hands of Magniac Smiths & Co., a London banking house. By agreement with Magniac Smiths, the George's Creek company might draw against the securities to the amount of £10,000, and the company did in fact draw a little more than $42,000.[10]

While Alexander was in London, Louis McLane, president of the Baltimore and Ohio Railroad, was also there negotiating with Baring Bros. for the sale of $3 million in Maryland state bonds issued to help finance the railroad. McLane also was only partially successful. The governor of Maryland commented: "For some time before the bonds were sent to England, the scarcity of money and the abundance of American securities, and their rapid fall in value, made it impossible to effect a sale or negotiate a loan on reasonable terms."[11]

August 31.—Another carpenter arrives from Baltimore with a letter from Mr. Wilson. He is assigned to Mr. Welch. Mr. T. Alexander leaves Lonaconing. Mr. Pauer and family also leave to spend some time in Frostburg on account of the sickness of Mr. P's child.

September 1, Sabbath.—Intelligence from Frostburg that Mr. Pauer had lost his [child] at 10 o'clock of the last night brought by a drunken Welshman, T. Lewis,

who delivers a note to Mr. Graham too late to make preparations for the child's interment until the morrow.

September 2.—Mr. Pauer buries his child at W. Shaw's cemetery. Pagenhart and Reuter are discharged from the carpenter's shop, and Pritchard and Hoffman from the smith's shop. D. Hopkins engaged in fitting blast pipes to cupola.

September 3.—Captain [John] Pickell [U.S. Engineers] and lady dine at Lonaconing.[12] The Scotch miners arrive at night.

September 5.—D. Hopkins reports the neglect and drunkenness of W. Hopkins and John Lewis. They are threatened with severe punishment, but the fines ($10 to each) are suspended on their promise that they will never go again to Buskirk's.

September 6.—C.B.S. visits Mr. Preuss to brush up his diligence in railroad matters. Mr. Preuss fails to keep an appointment with C.B.S. to determine a paper location of the railroad. The wives of David Price and John Davis (who have both been dismissed the company's service) are guilty of infamous conduct and are ordered to leave the place within one week upon pain of being forcibly removed.

September 7.—The morning occupied in instructions to Mr. Preuss and in locating upon paper the first mile of the railroad.

September 8, Sabbath.—Weisskettle's child (infant of 1 week) dies of a fit.

September 9.—Letter to Mr. Wilson on the subject of Mr. Duvall's account. $729 are still due to Duvall after he shall have cut with the shingle machine 20,000 shingles in one day. Pagenhart, who has been guilty of hunting on the Sabbath, is sent for. The "act" in relation to Sabbath breaking is read to him, and on his pleading ignorance and promising to do the like no more, he is excused. Clark [blacksmith] is again given the joint charge of the engine and blacksmith shop. Mr. Preuss and corps engaged in location of railroad.

September 10.—Colliers do not go to work at once, but after some conversation with C.B.S., and an endeavor to procure some promise of increased wages should they not be able to get more than 2 tons of coal per diem, they conclude to go to work at price announced for the date (50 cents per ton of 28 bushels).

September 11.—The little cupola essayed and performs well. Pipes are cast to complete the fixtures for blowing on three sides, in a succession of tuyeres. [As the molten metal collected at the bottom of the cupola furnace, the lower tuyere holes were plugged with sand or clay, and the blast was moved to higher holes.[13]]

September 12.—Directions given to Welch for the division of the upper rooms of the new store so as to give accommodation to the family of Mr. Owen on the east front, the doctor's office and the engineer office being on the north front second floor, and the engineers' quarters in the attic.

September 13.—Grauss and Gruber engaged on the foundation of the new smokehouse. The walls of the smokehouse will be 18" thick, the room 18 feet square on the interior and 15 feet to the eaves, thence to the ridge 6 feet. There are to be brick stoves on the outside at each end with their flues 9' square carried up on the wall 4½ feet and turned inwards, the smoke to escape as it can through the shingle roof. This construction is rendered necessary by the precarious condition of about

[9] Allegany Land Records, liber AA, folios 349–355. Although Alexander's statement of 1850 refers only to Magniac Smiths & Co., his correspondence makes it clear that some of the bonds were held by Magniac Jardine & Co., one of a group of London and Liverpool banking houses specializing in American loans. See J.H.A. to Magniac Jardine, Nov. 28, 1845, and Magniac Jardine to J.H.A., Jan. 3, 1846. J. H. Alexander Collection, the Maryland Room, Univ. of Maryland. See also Leland Hamilton Jenks, *The Migration of British Capital to 1875* (New York, 1938), p. 68, and fn. 13, p. 359.

[10] J.H.A. Statement of 1850, and C. E. Detmold, *George's Creek Coal and Iron Co.* (n.p., n.n., [1849]), p. 22.

[11] Message of William Grason to the Maryland legislature, Jan. 2, 1840, quoted in Hagerstown *Mail*, Jan. 10, 1840. For further comment on English coolness to American enterprise, see Walter B. Smith and Arthur H. Cole, *Fluctuations in American Business 1790–1860* (Cambridge, 1935), p. 42.

[12] In 1835 Pickell made surveys and examinations for the Maryland Mining Co. George W. Hughes, *Report of an Examina-*

tion of the Coal Measures . . . Belonging to the Maryland Mining Co. (Washington, 1836), pp. 3–8. In 1853 Pickell established his own coal mining company at Moscow, a few miles south of Lonaconing.

[13] Tomlinson, 1: pp. 344–345.

12,000 pounds of bacon now hanging in the cellar of the new store, and in which the worms already appear in numbers. This day is the one appointed for the removal of Mrs. D. Price and [Mrs.] J. Davis. The latter having taken herself off, the former and her chattels are conveyed beyond the company's boundaries. She takes shelter at "Burst keg's" [Buskirk's]. *Similis simili gaudet!* [Like takes pleasure in like.] Hill, the bailiff, calls on C.B.S. at midnight and informs that R[ichard] Williams, molder, had beaten his wife severely and that she had taken refuge in his house. He is told to keep her there till morning.

September 14.—E. Coombes is sent for, and upon the affidavit of R. Williams' wife, he is committed to gaol until the October court to answer his wife's complaint.

September 15.—R. Williams' wife is supplied with $50 from the wages due to him, to enable her to reach her friends in Dayton, Ohio, and departs this day in Rupert's stage.

September 16.—C.B.S. indisposed.

September 17.—C.B.S. confined to the house by gatherings in the ear. Rainy through the day, and freshet of the creek at night. The middle pier of the new footbridge back of Weisskettle's carried off, and one of the piers of the bridge near the powder magazine somewhat disturbed.

September 18.—The freshet has done some mischief to the race bank. Bricks cannot be hauled across the creek. Little cupola in blast and making tram plates.

September 19.—Richard Williams returns from Cumberland, having been bailed by Mr. Buskirk at the instance and on the security of Clark, the blacksmith, and Thomas Jones, molder. Steam engine does not perform well with 3 boilers. Mr. Hopkins is directed to blow with the whole 5 boilers.

September 20.—Steam engine not performing well, probably from bad firing. Clark is directed to examine the packing of the blast cylinder and finds it in good order. At 5½ P.M. the roof of the engine house takes fire and is either consumed or torn down by the hands, who promptly assemble at the ringing of the bell.[14] The fire completely extinguished before dark.

September 21.—Jarboe is directed to send three hands, and Mr. Welch to take his spare hands, and put up the roof of the engine house this day, tomorrow being Sunday. Mr. Schmidt is directed to send rough plank from sawmill. Welch accomplishes this job by night. Goodman Williams and D. Jones are employed in digging and stoning up a privy to the superintendent's house, which they undertake to make 12 feet deep and 6 × 4 feet on the inside, with dry walls 2 feet thick, for $24.

September 22, Sabbath.—No preaching in Lonaconing.

September 23.—Clark, the blacksmith, returns from Cumberland much disfigured by a fall from a horse (no doubt when intoxicated).

September 24.—Mr. Welch engaged in finishing the shell of the new church and framing the roof, and putting up shelves to the new store. Mr. Wilson is requested to advertise for a blacksmith and engineman, the former to secure $45 per month, and the latter $55 or $60, Clark having become too dissipated.

September 25.—Mr. Jarboe compelled to leave by the illness of his child. He receives a check for $100 and departs, leaving Mr. Davis in charge of his work.

September 26.—Mr. Preuss is detained at home by the sickness of his child.

September 27.—Mr. Preuss still detained from the field by the illness of his child. He sends to Cumberland for Dr. Wundsch.

September 28.—Dr. Wundsch arrives from Cumberland. Mr. Preuss' child is no better.

September 29, Sabbath.—No preaching. D. Jones absent in Ohio.

September 30.—Directions to Mr. Hill and W[illiam] Becker to go round and take down the names of those owning dogs, that the tax may be collected.

October 1.—Clark, the blacksmith, promises to reform and is retained upon trial. Richard Hughes, who had insulted Mr. Hill, the bailiff, is ineffectually sought after by the officers Stoub and Hill. They, however, find William Hopkins at Buskirk's in breach of his pledge to keep away. Memo: to exact the suspended fine of $10. Mr. Preuss buries his little child, which died last night at 12 o'clock, and addresses a letter to C.B.S. tendering his resignation of his appointment as topographical engineer to the company.[15] To this C.B.S. makes no immediate reply, thinking that Mr. Preuss, whose grief is very immoderate, is not in a state of mind to be reasoned with.

October 2.—This being election day, a general permission is given to voters to absent themselves for that purpose. Messrs. Harrison and H. Webster are sent with instructions to inform themselves what bearing the election will have upon the appointment of tax commissioners and to control the vote of the Lonaconing voters in obedience to the company's interests in that particular. In all other regards no intention is professed or entertained to restrict their rights as suffragans. Mr. Hill returns a list of 27 dogs for taxation.

October 3.—Mr. Preuss is persuaded by Mr. Pauer to alter his determination and requests leave of absence to the 1st of November, which is granted.

October 4.—Mr. Preuss sends in all the instruments, books, and papers connected with the railroad surveys [and] leaves for Baltimore. Welch puts up the rafters to new church, and cuts out the doors and windows.

October 6, Sabbath.—No preaching.

October 7.—Letter to Mr. Owen [who is to be rector of the company's church] and invitation to stay with C.B.S. until his arrangements may be completed. C.B.S. accompanies Mr. Serpell on the location of railroad. Great difficulty experienced to close with the experimental surveys. A strike occurs among Mr. Jarboe's workmen, of which C.B.S. is not informed, being entirely occupied with the railroad.

October 8.—Mr. Serpell again takes the field and fails totally to make the expected connection with the experimental survey.

October 9.—Dan Jones and G. Williams, the quarrymen, are employed by the day in making a yard to the little cupola and ways from the coke yard to it and also from the little cupola bridge house to the flat for the purpose of bringing up the pigs.

---

[14] ". . . on ringing the bell, we can muster a large [firefighting] force day or night with any number of buckets." Graham to Wilson, Oct. 23, 1839, GCC&I Co. Letter Book. A cistern in the upper part of the furnace building contained a large amount of water, and the creek was only 50 yards from the furnace.

[15] Apart from his grief, Preuss possibly felt that he would in any case be dismissed for defying Shaw's edict against consulting Dr. Wundsch. On the other hand, Preuss had for some time been dissatisfied with his job, which he described as "unbearable," and regretted having left the Coast Survey. Preuss to Ferdinand Hassler, Dec. 17, 1838, Hassler Papers, MS Div., N Y. Public Library.

October 10.—Mr. Serpell stays at home to protract his resurvey of Mr. Preuss' experimental line. His protraction closes with his own part of the location but cannot be made to do so with Mr. Preuss' work. Mr. Harrison is sent to repeat the same survey with a view to affirm or reverse the results of Mr. Serpell.

October 11.—Disputes between the families of J. Johnson, miner, and Morris, the butcher, settled by C.B.S. A presentment also on the part of Job Ray, miner, for criminal intercourse between his wife and Richard Richards, miner. Mr. Jarboe does not seem to be going on well. Many complaints against him by his hands. Weisskettle complains of being backbitten by Mr. Welch. Weisskettle's chief plasterer at the store very drunk. Letter from Mr. Wilson at night touching the relations between C.B.S. and Mr. T. S. Alexander.

October 12.—Mr. Thomas Alexander arrives on a short visit to the place. C.B.S.'s letter book is given to him on his departure for Cumberland, whither he departs at 4 P.M. to attend the approaching ejectment trial against Buskirk.

October 13, Sunday.—New complaints from J. Johnson against Morris, in consequence of which he is assigned another dwelling and removes at once, though it is the Sabbath, the object being to keep peace.

October 14.—Smokehouse finished. C.B.S. leaves Lonaconing for Cumberland. A riot occurs between Job Ray and Richards Senior and Junior. They mutually take out peace warrants against each other.[16]

October 15.—Information laid at Cumberland by C.B.S. against Buskirk for the riots at his house on the Sabbaths of the 4th and 18th [August]. The company's ejection suit also comes to trial, and at the moment of C.B.S.'s departure from Cumberland seems to be in a most favorable way.[17]

October 16.—Mr. Schmidt, Dr. Hermann, and other Germans go to Cumberland to declare their intention of becoming citizens.[18] C.B.S. returns to Lonaconing. Little cupola not in blast in consequence of the absence of the molder. Men employed in cleaning castings. [When the castings were removed from the mold, the runners which had carried in the hot iron were broken off, loose sand was scraped off with shovels and wire brushes, and the seams were smoothed with chisels and files.[19]] McNew, the new engineman, arrived on Monday during the absence of C.B.S. Mr. Serpell reports tonight that he had examined and remeasured the distances on the experimental survey for the railroad and finds the errors to be so great as to require a resurvey of the whole line.

October 17.—Mr. Graham concludes an agreement with Major Long for 200 barrels of flour at $5.50 and 200

bushels of white flint corn [at 50 cents[20]]. Mr. Pauer and Mr. Harrison directed to make sketches of the localities about the new store and sawmill for the purpose of furnishing to Mr. Wilson the information requisite to an insurance.

The correspondence with Richard Wilson concerning insurance furnishes a description of the buildings, a rough estimate of their value in their unfinished condition, and their approximate location.[21]

The store, 90 × 35 feet, had two floors, cellars, and an attic. At the end of October it had been shingled, the rooms had been given one coat of plaster, and three plasterers were at work. The carpenters still had to put in architraves, windows, and doors. Graham estimated that, except for painting, it would be finished by Christmas. Up to October 17, it had cost $3,700.

At the same date, the sawmill had cost about $12,600 (original estimate, $5,000). It had only a temporary roof, and of the three stories, only two—the pit and carriage floors—had been finished. The top floor, which was to be used as the pattern shop, was expected to be added about January 1, at which time the entire mill would be shingled and weatherboarded.

The superintendent's house had already cost $4,000, the amount allotted in the "memoir" of 1836. The lower floor was 50 feet square; the upper measured 50 × 38 feet. There was a front piazza 8 feet wide, and a rear one 4½ feet wide. The dwelling had been framed, shingled, and sheathed, and about half of the first coat of plaster had been applied. Flooring and doors were "got out roughly," and window and door frames, casings, and architraves were in progress. The extensive cellars were intended to contain a kitchen and a milk room. (It is interesting to note that the $200 paid for masonry in the cellars alone[22] equalled the cost of four houses for ordinary workmen.)

All three buildings were in the vicinity of the junction of Koontz Run and George's Creek.

October 18.—Letter from Mr. T. S. Alexander, desiring further information in regard to the riots at Buskirk's to be forwarded to the grand jury through Mr. Matthews. Letter to Mr. Matthews on that subject, and inquiry also made of him as to the award of the court in the company's favor, which it appears from report had dispossessed Buskirk of 100 acres and thrown upon him also the costs.[23] McNew engaged in packing the engine

---

[16] Job Ray appeared at the October session of the circuit court and requested continuance of the peace bond against Thomas and Richard Richards, swearing that he had cause to fear that they would "beat, wound, maim, kill, or do some other injury or hurt to his person." The court ordered the two Richards to post $100 bond to keep the peace. Clerk's Docket to Oct. Court, 1839, courthouse, Cumberland, Md.

[17] *John Guyer and John H. Alexander and Philip T. Tyson and the Georges [sic] Creek Coal and Iron Company* v. *Samuel Van Buskirk*. Action of Ejectment in Allegany County Court. Clerk's Dockets for 1838 and 1839, and Judgment Record R, folios 42–76, courthouse, Cumberland.

[18] See Appendix B for a list of George's Creek Coal and Iron Company employees who declared their intention of becoming citizens or were granted citizenship in 1839 and 1840.

[19] Tomlinson, 1: p. 345.

[20] Graham to Wilson, Oct. 18, 1839, GCC&I Co. Letter Book.

[21] Graham to Wilson, Oct. 17, 18, 19, and 23, 1839. *Ibid.* According to Scharf (*Western Maryland* 2: p. 1501), the store was destroyed by fire in the late 1870's. Local tradition has the company store at one time occupying part of the old "residency."

[22] Journal, June 20, 1839.

[23] The plat which was a part of the court record is now missing, and without it the jury's findings are no longer intelligible. On May 7, 1840, Samuel Van Buskirk agreed to sell to GCC&I Co. all the "lands enclosed within the lines of Com-

and the blast cylinder. John Lewis gone to Cumberland to take steps towards naturalization (civilization?).

October 19.—Mr. Baxter making graduation for tramway from the furnace and blacksmith shop to the mines.

October 20, Sabbath.—No preaching.

October 21.—Mr. Jarboe makes complaint against his journeyman, Davis, that he is undermining him and that he must discharge him. He is told to do as seems to him best. It now appears that the fire clay cannot be ground at the grist mill without ruin to the machinery, and that resort must be had to rolls after all. F. Germann, the painter, arrives from Baltimore, and a wagon with oils and paint. Germann is ordered to prime the cornices on the new house and afterwards with all diligence to paint the apartments intended for Mr. Owen.

October 22.—More difficulties with Mr. Jarboe about his hands. The prospect is unpromising for getting possession of the new house by Xmas. Plain talking to Jarboe about his habits.

October 23.—Letter from Mr. Owen at night. He will be at Lonaconing on Friday, the 1st November, and will preach on the following Sunday.

October 24.—Sundry complaints that D. James and J. Jones are giving annoyance to Job Ray about his wife. They are admonished. Peace is at length established between Jarboe and his men, and there is a prospect that the work will now go on better. C.B.S. is seized with a singular giddiness resembling epilepsy and [is] incapacitated for writing except a few lines at a time.

October 25.—C.B.S. still indisposed, and supposing that more exercise will be of service, rambles about the hills, but with no good effect. Vertigo and faintness continue.

October 26.—Mr. Preuss returns to Lonaconing.

October 27, Sunday.—This Sabbath, it is to be hoped, will be the last that will find the place without a preacher. Stork, the patternmaker, arrives.

October 28.—Letter from Mr. Preuss to C.B.S. about his surveys, and reply. C.B.S. still much indisposed. Mr. Serpell is directed to attend to the superintendent's duties. Strebig leaves Lonaconing, but is arrested by Stoub at the hest of a certain forsaken damsel and compelled to give bail for the maintenance of an heir expectant in the sum of $320.

October 29.—C.B.S. consults Dr. Hermann about his singular ailment and commences taking medicine. Stork is directed to fit up the lathe and to make patterns for rolls for the brick clay.

October 30.—Second letter from Mr. Preuss, and letter of instructions from C.B.S. to him.

October 31.—C.B.S. still very unwell. Mr. Serpell acting. Letter at night enclosing a long letter from Mr. T. S. Alexander. Mr. Preuss kicks at his instructions, but is persuaded by Mr. Graham to obey them.

November 1.—Letter to Mr. Wilson in reply to the letter and charges against C.B.S. on the part of Mr. T. S. Alexander. C.B.S. more indisposed.

November 2.—Mr. Graham estimates the October expenses at $7,000.[24] C.B.S. too much indisposed to attend to any business. Reverend Mr. Owen and family arrive at Lonaconing.

Owen, who seems to have been physically handicapped, was accompanied by his wife, three small children, and their nurse. They traveled by train to Frederick, where William Ferguson, a Cumberland liveryman, met them and took them to Lonaconing by carriage. Ferguson's usual charge for this trip was $30, but he may have reduced his fare in this instance.[25]

November 3, Sunday.—Morning service by Mr. Owen.

November 4 to 8.—In this interval nothing of great importance has occurred. C.B.S. again able to go about. Letter to Mr. Wilson in reply to one from him announcing the return from Europe of Mr. J.H.A. Jarboe has dismissed Davis, who is accordingly employed in the company's shop as chief carpenter and directed at once to make grate patterns. Stork [patternmaker] has not fitted up the lathe in consequence of there having been no turning tools sent with it. Mr. Graham orders some [from William Alexander].[26]

William consulted J.H.A., who told him to inform Graham that the tools were usually the private property of the patternmaker or were made by him "to suit his own notions." However, J.H.A. selected a set of chisels and gouges for wood-turning and arranged for the making of a few "implements of constant use in iron-turning." Any other tools could be made by the workman himself "out of the squared steel which is now on its way up."[27]

November 10, Sunday.—Service by Mr. Owen—congregation not very numerous.

November 11.—Mr. Baxter grading the way from the D level bridge to the blacksmith shop and furnace for the purpose of sending trams for repair. Letter to Mr. J.H.A. Number of boys and men now employed in the mines, 56. Number of tons of ore mined in October, 285.

November 11 and 12.—C.B.S. discovers that Mr. Preuss has disregarded his instructions and is going on in his own way. Letter to Mr. Preuss forbidding further operations and reply from Mr. P. declining (it is probable) to obey his instructions. As it appears from the messenger's account that the letter of C.B.S. has been totally disregarded by Mr. Preuss, the former concludes that his duty has been discharged and that there is no obligation upon him to expose himself to fresh impertinence from Mr. P. by the perusal of his letter. It is therefore handed to Mr. Graham with the seal unbroken for safekeeping until called for. Letter to Mr. J.H.A.

November 14.—Mr. Schmidt is sent to recall Mr. Preuss' hands, who come in.

November 16.—Organization of the church of St. Paul's, Lonaconing.

---

monwealth which were judged to be his" in this case, as well as lots 3720, 3721, 3722, 3772, 3773, and 3774 (50 acres each), and the tracts named "Skeleton," "Good Luck," and "Potato Hill" (amounting to 246⅝ acres.) Allegany Land Records, Liber AA, Folio 449.

[24] Estimates for the colliers, $321; ore miners, $1,753; furnace hands, $466; and laborers, $1,381; total, $3,921. However, since they took store accounts of $1,811, the men would get only $2,110 in cash at settlement. A sum of $500 was allotted to the contractors, and a payment of $1,937.70 to William Alexander for store goods. Graham to Wilson, Nov. 8, 1839, Letter Book. Graham's letter also indicates that since the last settlement he had drawn checks to the amount of $2,488.97, for which no detail is given.

[25] W. Alexander to Wm. Ferguson, Oct. 15 and 22, 1839. Letter Book 2, Welch and Alexander Record Books. W. A. described Owen as "bodily disabled."

[26] Graham to W. Alexander, Nov. 6, 1839. GCC&I Co. Letter Book.

[27] W. Alexander to Graham, Nov. 16, 1839. Letter Book 2, Welch and Alexander Record Books.

November 17, Sunday.—Service in the morning by Mr. Owen. Letter from Mr. J.H.A. requiring a quarterly tabular statement of Lonaconing statistics and a copy of the correspondence with Mr. Preuss.

November 18.—C.B.S. much indisposed.

November 19.—Letter to Mr. Alexander enclosing the original letters of Mr. Preuss and the duplicates of those of C.B.S. cut from the letter book. C.B.S. again indisposed with vertigo.

November 20.—Germann directed to pay [cover] the roofs of the store and the new house with coal tar.[28] Mr. Owen receives a letter from Mr. T. S. Alexander advising that the organization of the parish should be annulled, the name bestowed upon it not having been that intended—St. Peter's. Mr. Matthews and Mr. [Benjamin] Brown [surveyor] and the sheriff arrive to execute the writ of possession in the [Buskirk] ejectment case. C.B.S. too much indisposed to accompany them, deputes Mr. Serpell to that task.

November 21.—It is ascertained that the limestone quarry on the hill opposite Buskirk's sawmill becomes the company's property. Messrs. Matthews, Brown, and the sheriff depart.

November 22.—Mr. Baxter is directed to separate the ore and coal accounts; to keep a separate account of dead work, at least as much of it as relates properly to opening work, that it may if desired be charged under a separate head; and to prepare at the end of every month a statement of the number of workmen, their wages, and the results of their labor. These statements have heretofore very unnecessarily and improperly been left for the superintendent. Letter at night from Mr. Alexander disapproving of the correspondence between C.B.S. and Mr. Preuss, his dissatisfaction embracing both parties. Mr. Alexander directs the dismissal of the engineer corps and their servants. This day the proceedings of last Saturday were annulled and the church again organized as St. Peter's Parish. Letter to J.H.A. recommending Mr. Pauer for continued employment. Mr. Owen and family move to their residence this evening.

November 24, Sabbath.—Morning service by Mr. Owen. Very dangerous walking. But 6 persons present.

November 25.—The store roof is found to leak badly; it seems to be too flat. Mr. Owen and Mr. Graham depart for Baltimore [for church convention.[29]]

November 26.—Mr. Duvall makes his experiment with the shingle machine, which he has refitted. Duvall steams and cuts 24,000 shingles of small size.

November 27.—Grate patterns going on well. Stove pattern cast. Shingles counted, and Mr. Duvall receives from C.B.S. a certificate of the success of his machine. Letter at night from Mr. Alexander, commenting upon the quarterly statement. He also directs that the order

for dismissal of the engineer corps go into complete effect, reserving his decision in regard to Mr. Pauer's employment for the future.

November 28.—Mr. Schmidt is told to select the best of the engineer hands and to dismiss the same number from the laboring corps. The whole number of day laborers, including smiths, carpenters, drivers, and laborers is now reduced to 26.

November 30.—C.B.S. preparing drawing and specifications for the masonry of the buttress [for molding house] and cast iron centers for the same.

December 1, Sabbath.—No service in the morning. Welsh preaching in the afternoon by D. Jones.

December 2.—Stork is furnished the pattern of a new tram wheel and directed to suspend all other labor until it is made. Directions to Schwenner, who commences to lay the foundation of the buttress at molding house.

December 3.—The new church is now roofed, lathed upon the outside for the reception of shingles, and the sides about one-half covered. The window and door frames are also in.

December 4.—Mr. Graham and Mr. Owen arrive from Baltimore. Pattern of center for buttress arch finished and casts taken from it at $3\frac{1}{2}$ P.M.[30]

December 5.—Mr. Graham and C.B.S. preparing for the November settlement.[31] Mr. Baxter has finished measuring, and the yield for the month has been 346 tons, being $4.56 per ton, including all dead work. Difference in the price of ores between November and October $1.29 in favor of November.

December 6.—Measurement of stock of ore on hand—2,342 tons. Sawmill breaks a pinion belonging to the machinery for hauling in logs. Stork is directed to make a pattern for a new pinion. Letter from Mr. J.H.A., declining entirely Mr. Pauer's services, but as he promises to write to Mr. Pauer, C.B.S. determines to leave the unpleasant annunciation to himself. Mr. A censures the management of the ore department.

December 7.—Determine to give [Mr. Alexander] more mature views of the ore-getting, with which he seems to be dissatisfied. Letter at night from Mr. J.H.A. It is not opened, as it cannot be answered by the next mail, and its contents can only be productive of wakefulness and consequent headache to C.B.S.

December 9.—Mr. Alexander's letter contains a demand upon C.B.S. for his views of the ultimate prospects of the company, and suggests in delicate terms the possibility of C.B.S. being encumbered with assistance, which induces C.B.S. to acquaint Mr. Serpell that unless the railroad shall go on, his services will not be required by the company in the spring.

December 10.—D. Hopkins is informed of Mr. Alexander's wish that he shall labor as well as supervise furnace operations. He promises not to stand idle at a pinch, but does not distinctly promise to work a turn, giving as a reason for his unwillingness that he was not engaged for that purpose, and that it would prejudice his chance of employment as manager should he ever want a situation elsewhere, that he had here worked at day labor. Some changes are made today at the furnace. William Hopkins dismissed for drunkenness.

[28] Obtained from the gas light company in Baltimore. W. Alexander to Graham, Nov. 16, 1839. Letter Book 2, Welch and Alexander Record Books.

[29] Owen and Graham attended a special convention called by the Protestant Episcopal Church of Maryland to elect a bishop. On Nov. 28 they presented a petition from the vestry of St. Peter's Parish, asking to be received into the diocese. The petition was granted the same day. *Journal of a Special Convention of the Protestant Episcopal Church of Maryland (1839)* (Baltimore, 1839), pp. 10-11. It is interesting that the communion silver was presented to the parish by a British M.P., John Abel Smith (1801-1871), a partner in the family banking firm of Smith, Payne & Smiths and a stockholder in the George's Creek company (as was the banking firm). The gift is noted in Md. Historical Records Survey Project, *Inventory of the Church Archives of Md., Protestant Episcopal* (Baltimore, 1940), p. 209.

[30] Apparently it was Shaw's idea to use buttresses for supporting the molding house walls. Graham spoke disparagingly of Shaw's plan in a letter to Wilson dated April 3, 1840. GCC&I Co. Letter Book.

[31] Payments for ore and coal, $1,994.23; for furnace, $588.74; labor, hauling, and surveying, $1,351.73; contractors, $500; Wm. Alexander, $1,000; total $5,429.70. The amount which employees owed the company store was $1,800. Graham to Wilson, Dec. 7, 1839, Letter Book.

December 11.—Carpenters engaged in closing up the cast house and fitting in lights [windowpanes] to it.

December 13.—Directions to Mr. Graham to give no more money to Jarboe at present, and to keep Welch close.

December 15, Sabbath.—Letter from Mr. J.H.A. at night.

December 16.—Mr. Alexander's letter appears to be a general censure upon the management of C.B.S. and to include in one sweep every department of the concern—mines, furnace, sawmill, and buildings. Answer postponed until the contents shall have been duly pondered and digested.

December 17.—Stork, the patternmaker, dismissed. He is too slow.

December 19.—A wagon arrives from Baltimore bringing some patterns promised by Mr. Alexander of plough shares, corn shellers, etc. Mr. and Mrs. Graham depart to spend their Xmas in Baltimore.

December 20.—A new patternmaker, Phillips, engaged upon trial and set to work upon the grate patterns.

December 21.—Deep snow at night—2¾ feet on a level.

December 22, Sabbath.—Still snowing. Six inches more today. No mail.

December 23.—A rumor prevails that Mr. Alexander is coming. Rupprecht starts at 10 o'clock to break the way. Everything locked up about the place. Mr. Schmidt fits up a snow plough and makes a few paths. C.B.S. confined to the house with indisposition. Rupprecht returns at 4 P.M. Mr. Alexander is not with him.

December 24.—Nothing going on at cupola except the removal of snow from the cast house. Mr. Alexander arrives at 3 P.M.

December 25, Christmas.—Service by Mr. Owen.

December 26.—Reorganization of the laboring corps. All the laborers proper dismissed except George Blatter. Lists exhibited of the furnace and mine hands. Pay of laborers reduced to 87½ cents per day.

December 27.—Mr. Welch is informed that his work will conclude upon the completion of the store. Weisskettle is also told that no more plastering can go on until spring. Germann, the painter, is told that there is no no more work for him. T. Layson [whose leg was broken in a mine accident on January 23, 1839] is notified that his pension from the company will cease with the year.

December 28.—Mr. Alexander proposes many changes in the administration of the affairs of the place. Mr. Alexander resolves to stop Jarboe's job.

December 29, Sunday.—Path to church cleared out. Mr. Owen performs service but preaches no sermon. Still snowing.

December 30.—Everything buried in snow. Jones is directed to keep a moderate fire in the steam boiler to keep the connections from freezing. Sawmill hands directed to open the ways. Mr. Schmidt makes an attempt to plough; the plough breaks and is sent to the shop for repair. Hands at furnace and carpenters propping the engine house roof and shoveling off the snow from the furnace. Miners employed in removing snow from the ways.

December 31.—Rupprecht makes an ineffectual attempt to break the way to Frostburg through the snow. He can get no farther than H. Koontz' hill, where he can no longer see the fences. Jarboe is told to stop his job. C.B.S. aids him in arranging the items of his bill for unfinished work.

1840

Journal entries for 1840 end abruptly on February 8 without any hint of the sweeping changes which were to take place almost immediately thereafter. It is inconceivable that Shaw did not realize that his situation was untenable. Nevertheless, he continued to set down without comment the day-to-day events at Lonaconing, only once indulging in an exclamation mark to record a particularly poor performance at the cupola.

A few penciled notes follow Shaw's last entry. One of these refers to "Register 3 date 10 September 1840." If there was indeed a third volume of the Lonaconing journals, as this item implies, it disappeared many years ago. Scharf, writing in the 1870's, made no use of it in his *History of Western Maryland*. Since he drew heavily on the first two volumes, it may be assumed that, if the third volume had been available, he would have used it in continuing the history of Lonaconing and the George's Creek Coal and Iron Company. There is probably no chance that the missing volume will be found. The residual archives of this company were discarded and burned when the residency was converted to apartments in the 1950's.

January 1.—No work at furnace nor in mines. Davis, the carpenter, gets drunk and behaves with great indecency, for which Mr. Alexander directs his discharge. Rupprecht starts again with the mail and 2 horses (tandem) to break the way, not expecting to return until the morrow. Intensely cold at night.

January 2.—Sawmill, which froze fast, and all the machinery covered with a frostwork from the congealed steam. Molding sand frozen, and feed pipes to engine. Fires are kindled in stoves in the engine house, and the two temporary grates brought from the superintendent's house to thaw the sand in the cast house.

January 4.—Applications from W. Hopkins and John Williams for the remission of fines imposed by the superintendent for their misbehavior. Sawmill still frozen up. Cupola making tram plates.

January 5, Sabbath.—Service by Mr. Owen in his own house.

January 6.—Mr. Alexander gives orders to Mr. Baxter to reduce the mining force gradually to 50.

January 7.—Talk about reduction of prices at cupola.

January 9.—Mr. Jarboe presents his bill at night.

January 10.—Jarboe appears to have trumped charges against the company to the amount of nearly $1,000. Goods removed into new store.

January 11.—Mr. Alexander adjusts Mr. Jarboe's account nearly upon the basis of that made out by C.B.S. Jarboe is allowed by compromise about $200 more, but requests a month to decide whether he will accept the settlement.

January 12, Sunday.—Mr. Owen preaches again in his own house. Mr. and Mrs. Graham return to Lonaconing.

January 13.—Mr. Alexander preparing instructions for the administration of affairs until his return from Baltimore. Men at the cupola disposed to strike, their wages having been reduced 25 cents per day all around. No work in cupola.

January 14.—Patternmaker fitting up lathe and making patterns for wheat fans. Men at furnace conclude to go to work.

January 15.—Mr. Alexander departs for Baltimore, leaving memoranda for the superintendent's duties during his absence. Patternmaker employed upon plough pat-

FIG. 13. Casting from a cupola furnace. Tomlinson, *Cyclopaedia of Useful Arts.*

tern. Clark [blacksmith] directed to work in the cast house at cleaning and fitting up castings. McNew is directed to fire the engine and work it himself, and Myer to work in the molding house as spare hand and to assist McNew only when indispensable. Vogel discharged from blacksmith shop.

January 16.—Thermometer at 9 P.M. is 5° below zero of Fahrenheit. Sand frozen in cast house.

January 17.—McDonald directed to fit up a frame for suspending the 1,400-pound steelyard to weigh castings.

January 18.—Hill [bailiff] is directed to give notice to Ann Powell that she will be conveyed from the company's premises at 12 noon of the 20th if she be not sooner gone. Richard Williams reprimanded for his intimacy with her, and told that he will be expected to provide her with the means of traveling. No work in cast house except cleaning castings. Weisskettle relining cupola.

January 20.—Cupola again at work—2 bed plates molded and cast for corn shellers, and 13 boxes of tram plates. Mrs. Powell is sent from the premises at 11 o'clock in William Shaw's sleigh under escort of the bailiff. R. Williams and T. Beddoes go to Frostburg in Rupert's sleigh and insult Hill, the bailiff, at Frostburg.

January 21.—Dr. Hermann applies for work for the German miners who have been some weeks waiting, having understood Mr. Alexander that they were to be employed at the end of that time. Directions to Mr. Baxter to employ them until Mr. A's pleasure shall be understood. R. Williams, the molder, still absent without leave. No cast today.

January 22.—R. Williams still absent. D. Hopkins directed to inform him when he returns that there is no work for him. McDonald has finished the trestle for the weighing machine. Clark is directed to fit it up. Cupola makes 4 sets of the [wheat] fan mill irons and 6 boxes of tram plates.

January 23.—C.B.S. confined to the house with indisposition. Cupola making tram plates (15 boxes) and a set of fan mill irons. Memo: the weight of a set of fan mill irons is 23¼ pounds; that of the corn sheller and its parts, 151 pounds.

January 24.—C.B.S. severely indisposed. Made at cupola today 45 tram plates = 2,250 pounds; 1 set corn sheller irons; 1 set ditto fan mill; and other castings. Total about 2,520 pounds.

January 25.—C.B.S. unable to attend to business.

January 26, Sabbath.—Preaching by Mr. Owen in his parlor.

January 27.—Mr. Welch is employed in fitting up benches for the temporary use of the [store] ware room as a chapel. Upon attempting to start the engine, everything is found frozen fast. The supply and force pumps have to be disconnected and the ice removed from the seats of the valves, after which, upon a renewal of the attempt, the follower of the piston breaks in two, the piston being frozen fast to the cylinder. McNew, the engineman, attributes the casualty to the neglect of Hill, the bailiff, who had been directed he says by Mr. Alexander to keep up the fires in the engine house during the night. Hill denies having received such directions. C.B.S. still severely indisposed.

January 28.—C.B.S. sick. Richard Williams reappears at Lonaconing and is told to quit the premises by noon of tomorrow.

January 29.—Mr. and Mrs. Serpell leave Lonaconing for Baltimore. The crankshaft is broken at the gang saw gate. 1,246 pounds = work of cupola.

January 30.—Jones and his men busy in getting out the broken shaft. A new follower is cast for the piston. Thaw, and ice flood in the creek. The little footbridge at the mouth of Hill's Run carried away. The ice breaker and the exertion of Mr. Schmidt and hands save the new bridge near the powder magazine. Cast today = 21 tram plates and other fancy castings amounting to 440 pounds. Total, 1,490.

January 31.—J. Hill is reported to have behaved with great indecorum the preceding night, having abused and threatened McNew, the engineman, in the company's store. Inquiry is made into the matter by C.B.S., and Hill severely reprimanded for his intemperate behavior. Clark and D. Jones sent to assist in getting out the broken shaft at the sawmill, which [J.] Jones has not yet been able to do. J. Hill in the course of the day expresses a determination to leave the company's service at the expiration of a month. Tram plates and other castings amount to 1,289 pounds.

February 2, Sunday.—Mr. Owen preaches again in his own parlor.

February 3.—Patternmaker engaged in preparing shaft pattern. Castings today—fan mill irons and tram plates and wheels, 564¼ pounds!

February 4.—McDonald and Wagner making large crank wheel to drive the lathe in turning the shaft pattern and shaft for sawmill. Welch making benches for church service. Mr. Serpell returns to Lonaconing.

February 5.—Shaft cast.

February 6.—Patternmaker centers the new shaft for turning, but cannot go on for want of proper tools. Clark is ordered to make them. Letter from Mr. J.H.A. directing that the machinery already cast be sent down as samples by the first opportunity.

February 7.—Directions to carpenters to box up a fan mill, a corn sheller, and one of the ploughs. No cast today, Thomas Jones having lost his child. Mr. Baxter reports the average prices of ore-getting for the past month to be within 2 cents per ton the same as in the last month. Casting book examined, whence it appears that the molders are either very unskillful or that the cost of such machinery as ploughs, fans, and corn shellers will be 3 cents per pound at the cupola. The boxes with the patterns are sent up to Frostburg by Rupert to be forwarded to Baltimore.

February 8.—Thaw, and high freshet of creek. Wagner employed in fitting counter and desks in the office. Blacksmiths making tools for turning off sawmill shaft. Rupert goes with the mail on horseback. D. Thomas molding ploughs in cupola. T. Jones' child buried. Mr. Baxter's child is missing about dark, and much apprehension felt lest he should have gotten into the creek. He is found at 10 P.M., having strayed up to Wright's. [Journal ends]

END OF THE VENTURE

The drastic curtailment of operations at Lonaconing was only partly due to Alexander's dissatisfaction with the conduct of the works. A far more significant reason was that the entire Lonaconing venture had been predicated on access to market via the Chesapeake and Ohio canal, and in the summer of 1839 it seemed unlikely that the canal would arrive at Cumberland before the spring of 1842.[1] No advantage would accrue from producing large amounts of pig iron and castings which could reach the seaboard only at inordinate expense.[2]

Since connection with the canal was not immediately possible, the George's Creek company postponed construction of its railroad to the proposed terminal at Buck Lodge, and settled into a maintenance routine with reduced staff. As Shaw had earlier suggested, during this interval the cupola furnace would continue to cast materials for the company's use, particularly iron rails to be laid in the mines and on the tramroads connecting various parts of the works.[3]

Of the 260 workers employed by the company in the spring of 1839, only 78 remained in January 1840.[4] Furthermore, almost immediately after the last entry in the journal, Shaw either resigned or was dismissed, and left Lonaconing.[5] Graham became

superintendent, and it is his letter book which furnishes most of the information about activities between 1840 and 1845.

During 1840 laying of the tram rails proceeded as planned. The sawmill, besides cutting material for the company, produced a surplus of plank, shingles, laths, and scantling for sale to the public. In the first nine months of the year the cupola cast tram plates, tram wheels, vault doors for the company store, and agricultural implements. In October iron operations were suspended; the cast house was torn down; and construction of the permanent molding house was resumed.

Judging from the value of goods sent up by William Alexander, store business fell off sharply. William's invoices for 1839 amounted to $16,400; for 1840, $6,500; for 1841, $2,500.[6] In May 1840 the company announced a policy of paying wages three-fourths in store orders and one-fourth in cash,[7] but even this requirement did not compensate for the loss of customers resulting from the exodus of 70 per cent of the workmen and their families.

The reduction in force also affected the development of St. Peter's parish. Work on the church stopped, and diocesan records do not show that the building was ever consecrated as a place of worship. In November 1840, shortly after the Reverend William Whittingham became bishop of the Maryland diocese, he visited his western parishes and preached to a small congregation in "the temporary chapel of St. Peter's parish, Lonaconing, after morning prayer by Mr. Owen."[8] During the summer of 1840 Owen had several times been invited by the citizens of Frostburg to organize a parish there, and had finally accepted. At the time of the bishop's visit, Owen was living in Frostburg, serving as rector and teacher at an annual salary of $300 plus a dwelling house and school building.[9] Presumably Owen had intended also to minister to the small Lonaconing congregation, but he wrote to the bishop on January 1, 1841: "My connexion with St. Peter's church, Lonaconing, as its Rector, has just terminated in consequence of the Decision of the Board of Directors, that 'non-residence is a vacation of the Contract.'"[10] After 1841 no mention of St. Peter's appears in the Journals of church conventions until 1859, when the present building was begun on a company-donated lot near the north end of town.[11]

[1] The canal company announced in 1837 that the contract for the final section between the Cacapon tunnel and Cumberland would be let in August of that year, and predicted completion early in 1840. C&O Canal Co., Ninth Annual Report, June 12, 1837 (Washington, 1837), pp. 11, 14. In 1839, with 50 miles still to be constructed, the president and directors of the canal company reported that the project would be finished by Oct. 1, 1841. Ibid., Eleventh Annual Report, June 3, 1839 (Washington, 1839), pp. 9–10. GCC&I Co. felt that in view of the "unprecedented money crisis" the spring of 1842 was a more probable completion date. GCC&I Co. Report 1839, p. 10. As it turned out, the canal was completed in 1850.

[2] Wagoners charged 50 cents per hundred pounds (amounting to $10 per short ton) to take iron from Lonaconing to Baltimore. W. Alexander to Graham, June 13, 1839, Letter Book 2, Welch and Alexander Record Books. Contrast this with the freight rates on the canal: $2.50 per ton for pig iron and $3.50 per ton on castings carried from Williamsport to Georgetown. Adv. Hagerstown Mail, Mar. 10, 1837. Pig iron could be sent from Williamsport to Baltimore via canal and railroad for $4.35 per ton. Ibid.

[3] Journal, July 26 and Aug. 3, 1839.

[4] Graham to W. Alexander, Apr. 26, 1840. Letter Book.

[5] Graham to R. Wilson, Feb. 11, 1840. Letter Book.

[6] Totals computed from entries in acct. book, Welch and Alexander Record Books.

[7] Graham to Wilson, Apr. 17, 1840. Letter Book.

[8] Journal of a Convention of the Protestant Episcopal Church of Md. 1841 (Baltimore, 1841), p. 19.

[9] Ibid., and Owen to Whittingham, Jan. 1, 1841, Md. Diocesan Archives, on deposit with Md. Hist. Soc.

[10] Diocesan Archives.

[11] Md. Historical Records Survey Project, Inventory of the Church Archives of Maryland, Protestant Episcopal (Baltimore, 1940), p. 208; and Allegany Land Records, liber 18, folio 335.

It appears that the company's physician, unlike its pastor, found it possible to make a living at Lonaconing. Whether or not still under contract, Dr. Hermann, who had married one of David Hopkins' daughters, stayed there until late in 1846, when he moved to Cumberland.[12]

Graham's letters suggest that the cupola was out of blast during all of 1841 and possibly much of 1842. The company's dwindling income from Lonaconing came from occasional sales of pig iron to Cumberland foundries and the sale of lumber to the Mount Savage iron works, then being erected. Selling to local residents as well as to employees, the store managed to keep going by limiting its stock and practicing the most rigid economies.[13] In the fall of 1842, expecting soon to put the high furnace into blast, Graham offered the Baltimore and Ohio Railroad, which had finally arrived at Cumberland, "any quantity" of pig iron at $25 a ton.[14] From this point on, the letter book becomes less informative. Most of the correspondence deals with orders for the store. There are no letters at all between August 28 and December 23, 1843, and none between December 29, 1843, and March 18, 1844.

In 1844 the company made several unsuccessful attempts to operate the blast furnace with charcoal instead of coke or coal.[15] Discouraged by this failure following several years of relative inactivity at Lonaconing, J. H. Alexander asked Christian E. Detmold, a civil engineer, to visit Lonaconing and make suggestions for the future operation of the enterprise. Detmold, who was one of the original stockholders and directors of the George's Creek Coal and Iron Company, had recently become interested in experimenting with the German method of using furnace gases for heating the blast in the manufacture of iron.[16] As a result of the consultation, the company leased to Detmold for a period of seven years the blast furnace, cupola, and steam engine, together with the right to mine raw materials, cut timber, build roads, and erect whatever buildings he needed for housing his workmen or for carrying on the manufacture of iron.[17]

When Detmold signed the lease, he understood that the company would try to raise money for a

FIG. 14. Blast furnace, bridge house, and cast house of a typical antebellum iron works. Overman, *Treatise on Metallurgy.*

railroad to connect Lonaconing with the railroad which the Maryland Mining Company was building to meet the Baltimore and Ohio at Cumberland.[18] To finance its railroad, the George's Creek company issued $150,000 in bonds secured by the mortgage of Commonwealth to William Bard and Edward Whitehouse of New York City and John H. Alexander of Baltimore.[19] Although the loan was fully subscribed, the cash proceeds were substantially diminished by payments for other purposes, and the remainder was not sufficient for construction of the line.[20] After making surveys and purchasing some small amounts of land for right of way, the company made no further effort to provide transportation between Lonaconing and Cumberland.[21]

In the meantime, Detmold moved to Lonaconing and began to put the works in readiness for blowing in. Since the plant had been without informed resident superintendence for more than five years, the furnace and mines required a great deal of work before operations could begin. As part of the renovation, Detmold overhauled all the boilers and put up one new one, installed a new fly wheel, shaft, cylinder, pump, piston, and cog wheels in the steam engine, laid new steam pipes, and practically rebuilt the engine house.[22] In spite of delays caused by the extensive repairs and the unusual severity of the winter, Detmold blew in the furnace in May 1846.[23] By 1847 he was doing a flourishing business, "sending to the East some of the best Iron manufactured in the United States."[24]

Almost immediately after Detmold had demonstrated that under skilled management the furnace could make a profit, the George's Creek Coal and Iron Company began prolonged legal action, first (unsuccessfully) to evict him, and then to interfere

[12] Cumberland *Alleganian*, June 5, 1847. Hermann's advertisement is dated Nov. 21, 1846. Later he returned to practice in Baltimore. In 1859 he emigrated to Oregon, taking a party of settlers from Baltimore and Philadelphia and at least one family from Lonaconing. George Bennett, "History of Bandon," *Oregon Hist. Quart.* 28: pp. 334–341.

[13] For example, Graham instructed Wm. Alexander to send receipts and invoices by wagon in order to save on postage. Graham to W. Alexander, Aug. 3, 1841, Letter Book.

[14] Graham to B&O, Oct. 4, 1842, Letter Book.

[15] C. E. Detmold, *George's Creek Coal and Iron Co.* (n.p. [1849]), pp. 2–3.

[16] *Ibid.*, and James M. Swank, *History of the Manufacture of Iron in All Ages* (2d ed., Philadelphia, 1892), pp. 454–455.

[17] Detmold, Appendix A.

[18] *Ibid.*, p. 3.

[19] Allegany Land Records, liber 1, folio 448; Detmold, p. 4; and Magniac, Jardine & Co. correspondence, J. H. Alexander Collection, Maryland Room, Univ. of Md.

[20] Detmold, pp. 5–6, 22.

[21] *Ibid.*, pp. 6–7; and Hazlehurst correspondence, J. H. Alexander Collection.

[22] Testimony of David Hopkins in *GCC&I Co.* v. *C. E. Detmold*, Chancery Records No. 8284, Md. Hall of Records.

[23] Detmold, p. 6.

[24] Cumberland *Civilian*, Nov. 19, 1847.

with his operations.[25] Nevertheless, he not only kept the furnace in blast, but also built 8½ miles of tramroad from Lonaconing to Clarysville, there connecting with the Maryland Mining Company's railroad.[26] Thus he was able to market his product and to bring in iron ore from other areas when supplies at Lonaconing ran low.[27] By 1850 Detmold had $200,000 invested in the works. Offsetting this tied-up capital was an annual production of 2,500 tons of pig iron valued at $50,000.[28]

Since 1844 the George's Creek company had become increasingly interested in developing its coal resources. The unsuccessful railroad negotiations of 1845–1846 had in fact contemplated a line connecting one of the company's coal mines with the Maryland Mining Company's railroad. Only incidentally would a spur from the mine to Lonaconing provide an outlet for Detmold's iron.[29] The abandonment of the project, although a set-back for Detmold, was in the long run a good thing for the company: there still remained the possibility of a direct connection with the Baltimore and Ohio Railroad, depending upon the route chosen for its westward extension from Cumberland.

Largely because of J. H. Alexander's glowing forecasts of the coal trade,[30] the Baltimore and Ohio decided to push southwest from Cumberland twenty-one miles along the Maryland side of the North Branch of the Potomac, thence over an iron bridge to the Virginia shore of the river, terminating at Piedmont, opposite Westernport and the mouth of George's Creek. Alexander's intent was clear. His company, with a well-established operating base, would exercise its charter right to build a railroad "to some convenient point . . . near the mouth of George's Creek" and would control the transportation of coal along the entire valley from Lonaconing to the Piedmont depot.

When the Baltimore and Ohio announced its decision, the George's Creek Coal and Iron Company appointed William H. Smith as engineer and super-intendent, replacing Graham.[31] By the autumn of 1850 Smith had completed surveys and locations for a railroad from Lonaconing to Westernport.[32] The Baltimore and Ohio reached Piedmont in July 1851.[33] In September the George's Creek company started work on the southern end of its line, beginning with the bridge which would take the railroad across the Potomac from Westernport to Piedmont.[34]

It would be interesting to know the circumstances which made it possible for the corporation in 1851 to undertake this sort of project, which it could not afford in 1846. No financial statements or books of account have come to light to show how much cash the company had on hand when Smith first put a crew of one hundred to work at Westernport. The amount must have been substantial, because the directors did not find it necessary to negotiate a new loan in advance of construction. In July 1852 they did issue $170,000 in bonds secured by a mortgage on Beatty's Plains.[35] J. H. Alexander purchased more than $40,000, and several other original stockholders bought substantial amounts, of these bonds.[36] As the railroad neared completion, anticipation of "immense" profits sent the price of George's Creek shares soaring on the Baltimore market.[37]

The new bond issue more than paid the cost of the railroad and equipment. An increase of $250,000 in the company's capital stock, to be issued as needed, provided an additional source of funds for developing coal production.[38] During part of 1852, the company mined no coal[39] and therefore had little on hand when its railroad opened for business on May 9, 1853.[40] The Cumberland *Miners' Journal*, which published weekly statistics of the coal trade, began including figures for the George's Creek shippers in the issue of June 24, 1853. For the week ending June 18 the George's Creek company railroad carried 1,061 tons, but only 351 of these came from its own mines. Soon afterward the company contracted to supply coal to the United States government and to the Cunard steamship lines, and was reportedly

[25] Detmold, pp. 7–9. Allegany County court dockets indicate a series of suits, postponements, trials, and appeals beginning in Oct. 1846 and extending to Oct. 1852.

[26] Cumberland *Civilian*, June 22 and Nov. 23, 1849. GCC&I Co. unsuccessfully sued in chancery court to enjoin construction of the tramroad. *Ibid.*, Jan. 5, 1849; and *GCC&I Co. v. C. E. Detmold*, Chancery Records No. 8284, Md. Hall of Records.

[27] Testimony of David Hopkins, *GCC&I Co. v. C. E. Detmold*, Chancery Records No. 8284; and Cumberland *Civilian*, Nov. 23, 1849.

[28] MS. census, Md., "Products of Industry," 1850. Md. State Library.

[29] Gonder, Hazlehurst & Co. to R. Graham, July 29, 1845, Letter Book.

[30] Alexander claimed credit for the B&O decision. J. H. Alexander's Statement of 1850, pp. 2–3, Alexander Papers.

[31] Cumberland *Civilian*, June 29, 1849. The company's interest in the coal trade was demonstrated by one of Smith's first official acts—participation in a convention at which western Maryland coal operators agreed on setting prices and wages. *Ibid.*

[32] *Ibid.*, Oct. 11, 1850.

[33] *Ibid.*, July 18, 1851.

[34] Cumberland *Miners' Journal*, Sept. 26, 1851.

[35] Allegany Land Records, liber 8, folio 662.

[36] Hammer Acct. Books, vol. 1, folio 98. See also Dudley Selden to J. H. Alexander, July 17 and Nov. 7, 1851, Dec. 4, 1852, and Feb. 7, 1853; and Wm. Alexander to J.H.A., Jan. 25, Feb. 4, and July 6, 1853, Alexander Papers. Selden's letters concerned Charles Oliver's purchases as well as his own.

[37] Cumberland *Miners' Journal*, Jan. 28, 1853.

[38] Md. *Laws* 1853, ch. 25.

[39] Cumberland *Miners' Journal*, July 30, 1852.

[40] *Ibid.*, May 13, 1853.

organized for full production.[41] In spite of a miners' strike which lasted from early January to the end of March,[42] shipments for the Piedmont region (consisting of the seven companies along lower George's Creek) during 1854 amounted to more than 181,000 tons.[43] Over 30 per cent came from George's Creek Coal and Iron Company mines. In 1855 the railroad carried nearly 225,000 tons, of which more than 39,000 were from George's Creek Coal and Iron Company mines.[44] In 1856, with a contract to supply coal to the Baltimore and Ohio Railroad, the company on one occasion sent down a single train of 102 loaded cars.[45] During this year George's Creek Coal and Iron was the biggest producer on its line, shipping over 100,000 tons out of a total of nearly 267,000.[46]

In September 1856 the company extended its railroad two miles northward and connected with the Cumberland and Pennsylvania Railroad (owned by the Mount Savage Iron Company). A through line from Cumberland via Mount Savage and Frostburg to Lonaconing[47] now made it possible for the George's Creek mines to ship on the Chesapeake and Ohio canal as well as on the Baltimore and Ohio Railroad.

The depressed condition of the iron trade undoubtedly was the major reason for the company's decision to relegate the furnace to a minor part in its operations. The tariff bill passed by Congress in 1846 removed much of the protection which had enabled American ironmasters to meet the price of English rails and other manufactured iron products.[48] Attributing their difficulties to the new low duties, Americans closed their rolling mills and blew out their furnaces.[49] In western Maryland financial difficulties forced the Maryland and New York Iron and Coal Company and two smaller companies out of business. The Maryland and New York Company works at Mount Savage, with a complex of 280 workers' houses, blast furnaces, rolling mill, cupola, puddling and heating furnaces, and nine miles of railroad, were sold for $215,000, a fraction of their cost.[50] Nevertheless, while misfortune was over-

FIG. 15. Remains of the Lonaconing furnace, *ca.* 1911. The furnace was entered on the National Register of Historic Places in 1973. Maryland Geological Survey.

taking the iron industry, there was "a good deal doing in coal." [51]

When Detmold's lease expired in October 1852, the George's Creek Coal and Iron Company did not attempt to operate its iron works and seemed reluctant to let anyone else do so. In response to Tyson's offer to lease the furnace, Thomas Alexander wrote, "I will not agree in any manner to embarrass our operations for the sake of the furnace." [52] Whether or not the company did make some arrangement with Tyson, the furnace was again put in blast in August 1854 "after a suspension of several years." [53] Using ore from newly-discovered sources, it began turning out about sixty tons a week, "being at the rate of 3,120 tons per annum." [54] The blast continued into 1855, when the furnace produced 1,860 tons before it was blown out for the last time.[55]

From a twentieth-century vantage point it is clear that the "bad news from England" which called a halt to expansion in 1839 in fact forestalled utter disaster for the George's Creek Coal and Iron Company. Had Alexander's plans for four furnaces,

[41] *Ibid.*, July 29, 1853; and Thos. Alexander to J. H. Alexander, Aug. 7, 1853, Alexander Papers.

[42] Katherine A. Harvey, *The Best-Dressed Miners* (Ithaca, 1969), pp. 140–143.

[43] Robert G. Rankin, *The Economic Value of the Semi-Bituminous Coal of the Cumberland Coal Basin* (New York, 1855), table following p. 62.

[44] Cumberland *Telegraph*, Jan. 3, 1856.

[45] *Ibid.*, Jan. 3 and July 17, 1856.

[46] *Ibid.*, Jan. 1, 1857.

[47] *Ibid.*, Sept. 4 and 25, 1856, and Dec. 17, 1857.

[48] Baltimore *American*, Oct. 24, 1845.

[49] Cumberland *Civilian*, Aug. 11, 1848, quoting New York *Express;* and Baltimore *American*, Sept. 7, 1849.

[50] Cumberland *Alleganian*, Aug. 21 and Nov. 13, 1847. Lena furnace on the outskirts of Cumberland and Vulcan furnace in Hampshire County, Virginia, were sold in 1848 and 1849, respectively. Cumberland *Civilian*, Nov. 17, 1848, and Jan. 12, 1849.

[51] Cumberland *Civilian*, June 1, 1849, quoting Horace Greeley's account of his recent trip to the Cumberland area.

[52] T. S. Alexander to J. H. Alexander, Jan. 26, 1853, Alexander Papers.

[53] Baltimore *Sun*, Aug. 7 and 12, 1854.

[54] *Ibid.*, Oct. 27, 1854.

[55] Joseph T. Singewald, Jr., *Report on the Iron Ores of Maryland* (Baltimore, 1911), p. 142. The date of blowing out is given as 1856 by James W. Thomas and T. J. C. Williams, *History of Allegany County* (n.p., 1923) 1: p. 538. Possibly the last of the Lonaconing iron was sold in midsummer of 1856, when a Baltimore commission merchant wrote to a prospective customer: "This Furnace has gone out of blast and we offer it ['about 120 Tons of No 1 Lonaconing dark Grey Iron and about 30 Tons No 2 Iron'] to you at these low prices to close it off at once. The furnace will not again go into blast as it does not pay them." Stickney & Co. to T. Chaffee, July 11, 1856. Builders Iron Foundry Papers, Manuscripts and Archives Dept., Baker Library, Harvard Graduate School of Business Administration.

a rolling mill, and other accessory buildings been carried out, there is no reason to suppose that their operation would have resulted in anything but failure. The Mount Savage Iron Works, under their new and presumably knowledgeable owners,[56] enjoyed a few years of prosperity before the Civil War,[57] but did not survive the war years. In 1864 the Consolidation Coal Company took over the Mount Savage properties.[58] In contrast, the George's Creek Coal and Iron Company remained in business until 1910, and its successor until 1952.

Both the George's Creek and the Mount Savage ventures were at a disadvantage because they were technologically too far advanced for their time and location. Small charcoal furnaces survived on the frontier because they produced for local use the fine quality iron demanded by blacksmiths and small foundries. Iron smelted with coke or raw coal was particularly suited for rails and heavy industrial castings, but, unless their product could be delivered to market cheaply, iron works using the new smelting process had no hope of survival. It was precisely because of the network of anthracite canals and railroads that hard coal, rather than bituminous, captured the iron market after 1855.

The fact that western Maryland ore supplies were less than estimated contributed to the abandonment of iron manufacture in this region. However, if there had been adequate transportation, ore could profitably have been brought to the superb supply of George's Creek coal. The Pennsylvania Railroad's connecting line did not reach Cumberland until 1872, too late to reverse the total commitment to coal mining by all the companies in the area.

In terms of American industrial history, the George's Creek Coal and Iron Company must be given credit for its pioneering work in the use of coke and of raw bituminous coal. It must also be acknowledged that its initial success prompted the Mount Savage owners to complete their plant and enabled them to become the first American producers of rolled rails. The George's Creek success also influenced other iron manufacturers to change from charcoal to coke and to build new furnaces expressly for the use of coke. All of these new furnaces built before the Civil War followed Alexander's design.[59] Thus the pattern and the practice for bituminous smelting were established, but the coke furnaces could not compete against the anthracite furnaces east of the Alleghany Mountains.[60] Coke came into its own only in the 1860's, when Pittsburgh began its rise to preeminence in the manufacture of iron and steel.

---

[56] Erastus Corning, John F. Winslow, and John Murray Forbes headed the company which bought the Mount Savage plant in 1847. Cumberland *Civilian*, Nov. 12, 1847.

[57] For a brief history of this company, see Jay Douglas Allen, "The Mount Savage Iron Works, Mount Savage, Maryland. A Case Study in Pre-Civil War Industrial Development." Master's thesis, Univ. of Md., 1970.

[58] Charles E. Beachley, *History of the Consolidation Coal Company 1864-1934* (New York, 1934), pp. 17-19.

[59] Frederick Overman, *The Manufacture of Iron in all its Various Branches* (Philadelphia, 1850), p. 175. Still in fairly good condition, the prototype at Lonaconing was entered on the National Register of Historic Places in 1973.

[60] Overman, p. 174.

APPENDIX

A. PASSENGER LIST OF BARQUE *TIBERIAS*

District of Baltimore

Port of Baltimore, Sept. 10—1838

I, George Sears, do solemnly and truly swear that the within list Subscribed with my name contaigns to the best of my Knowledge & belief a just and true account or report of all passengers who have been taken on board the Barque Tiberias at New Port (Wales) or at any other foreign port or at Sea & brought in said vessel into any district of the U. States since the departure from said port of New Port. Sworn the 11 Sept. 1838.

The whole of the within list of passengers are destined for George's Creek Co except Mary Bannista who is for this city.

Sept. 11, 1838

A List of Passengers on board of the American Barque *Tiberias* of Boston U.S. 299 Tons Reg^r. George Sears master bound from Newport to Baltimore U.S. Destination is George's Creek Co. for all except Mary Bannista—Baltimore.

B. NATURALIZATION RECORD OF GEORGE'S CREEK COAL AND IRON COMPANY EMPLOYEES, 1839–1840

The clerk's dockets of the Allegany County Circuit Court show that in 1839 and 1840 the following employees of the George's Creek Coal and Iron Company declared their intention of becoming citizens of the United States. The clerk's spelling is retained throughout.

Name	Age	Occupation
Mary Bannista	54	Lady
Thomas Phillips	26	Collier
Margaret Phillips	26	—
Mary Phillips	5	—
Cecil Phillips	3	—
William Phillips	infant	—
John Johnson	25	Collier
Elizabeth Johnson	22	—
John Johnson	infant	—
Herbert Watkins	29	Collier
Jane Watkins	29	—
Jane Watkins	5	—
Mary Watkins	2½	—
Rees Rees	50	Collier
Margaret Rees	50	—
John Rees	25	Collier
Thomas Rees	23	Collier
Jenkins Rees	21	Collier
Daniel Rees	19	Collier
Roger Williams	42	Collier
Charlotte Williams	33	—
William Williams	12	Collier
Roger Williams	10	Collier
Margaret Williams	7	—
Elizabeth Williams	4	—
Charlotte Williams	infant	—
Benjamin Thomas	45	Collier
Hannah Thomas	40	—
John Thomas	23	Collier
James Thomas	20	Collier
Benjamin Thomas	18	Collier
William Thomas	15	Collier
Deanna Thomas	6	—
Joseph Thomas	3	—
Phillip Thomas	2	—
Jane Thomas	infant	—
William Davies	40	Collier
Ann Davies	40	—
William Davies	16	Collier

Name	Age	Occupation
Sarah Davies	13	Collier
John Davies	11	—
Ann Davies	9	—
Isaac Davies	6	—
George Treasure	42	Collier
Elizabeth Treasure	40	—
William Treasure	19	Collier
Job Treasure	14	Collier
Ann Treasure	12	—
Susannah Treasure	10	—
John Treasure	4	—
Hannah Treasure	3	—
Caroline Treasure	infant	—
John Williams	31	Collier
Maria Williams	33	—
Adam Williams	15	Collier
Abraham Williams	13	Collier
John Williams	11	Collier
Ann Williams	8	—
Maria Williams	7	—
Benjamin Williams	2	—
John Lewis	36	Founder
Ann Lewis	37	—
Margaret Lewis	16	—
Hannah Lewis	9	—
Elizabeth Lewis	7	—
Ann Lewis	4	—
Thomas Lewis	38	Founder
John James	22	Collier
David James	16	Collier
Gwenelly Davies	24	—
Elizabeth Davies	22	—
Jenkin Thomas	35	Farmer
Mary Thomas	35	—
David Thomas	3	—
Margaret Thomas	infant	—
Richard Richards	23	Collier

Produced at the Custom House Newport this 31 June [*sic*] 1838

April Term 1839

German: John Hermann Dittmer and Anthony Fink
Welsh: Levi Harris and John Williams

October Term 1839

English: George Treasure
German: George Blatter, John Ditmer, John Leonard Fratz, John Grimes, Mathias Gruber, John Schwenner, Frederick Smith, Augustus Weiskettle, and Edward Zachariah.
Welsh: William Davis, David Hopkins, William Hopkins, Daniel James, John Johnson, John Jones, Edward Lewis, John Lewis, Thomas Phillips, Thomas Probert, John Reese, Benjamin Thomas, Herbert Watkins, and Roger Williams.

April Term 1840

German: Anthony Justus

October Term 1840

Welsh: William James and John Jones

The clerk's records also show that the following were granted citizenship: April Term 1839, Michael Milholland, Irish; October Term 1839, Christian Blatter, Jr., German; October Term 1840, [Dr.] Henry Hermann, German, and Thomas Layson, Welsh.

BIBLIOGRAPHY

I have listed below only the materials on which I have based my introduction. concluding chapter. and editorial notes. From roughly contemporary authorities on iron metallurgy and manufacture (largely English), I have tried as far as possible to draw on the American experience in describing construction and processes.

In identifying persons mentioned in the journal, I have cited some sources so obvious that it might seem unnecessary to mention them except for purposes of evaluation. It should be noted that a number of scholars have questioned the accuracy of Appleton's *Cyclopaedia*, Scoville's *Old Merchants of New York*, and the various Moses Beach lists of "wealthy citizens." Brief identifications of Allegany County residents are bracketed in the text, and their source, the manuscript population census, is not given.

MANUSCRIPTS

Allegany County, Maryland (Courthouse, Cumberland)

Assessors' Records 1837–1841
Dockets of Clerk of Circuit Court
Judgment Records of Circuit Court
Land Records
Minutes of County Commissioners 1837–1841

Eleutherian Mills Historical Library

Map of Frostburg, Maryland, Coal Field

Enoch Pratt Free Library, Baltimore: Maryland Dept.

J. H. Alexander's map of Allegany County

Harvard University, Baker Library, Graduate School of Business Administration

Builders Iron Foundry Papers

Maryland Academy of Science, Baltimore

Minutes of the Maryland Academy of Science and Literature 1835–1844

Maryland Hall of Records, Annapolis

Adjutant General's Records
Certificates of Survey, Allegany County
Chancery Records
Land Records, Allegany County (microfilm)

Maryland Historical Society, Baltimore

Alexander Papers
George's Creek Coal and Iron Company Letter Book
Hammer Account Books
Tyson Papers
Tyson Record Book
Welch and Alexander Record Books
On deposit with the Society: Maryland Diocesan Archives, Protestant Episcopal Church.

Maryland State Library, Annapolis

Manuscript census, Maryland, "Products of Industry" 1850

National Archives

Manuscript Population Census, Allegany County, Maryland, 1830, 1840
Ships' Passenger Lists
Post Office Department Records
Records of the Chesapeake and Ohio Canal Company

New York Public Library, Manuscript Division

Hassler Papers (Astor, Lenox and Tilden Foundations)

University of Maryland, Maryland Room, McKeldin Library, College Park Campus

John H. Alexander Collection

Virginia State Library, Archives Division, Richmond

Records of the Board of Public Works

NEWSPAPERS AND PERIODICALS

American Journal of Science and Arts
Baltimore *American*
Baltimore *Sun*
Baltimore *Bee*
Cumberland [Md.] *Alleganian*
Cumberland *Civilian*
Cumberland *Miners' Journal*
Cumberland *Telegraph*
Hagerstown [Md.] *Mail*
Harper's Monthly
Harper's Weekly
Hunt's *Merchants' Magazine*
Journal of the Franklin Institute
Mechanics' Magazine
Oregon Historical Quarterly
Portland [Me.] *Advertiser*
Pottsville [Pa.] *Miners' Journal*

BOOKS AND ARTICLES

ABBOTT, COLLAMER L. 1965. "Isaac Tyson. Jr.. Pioneer Mining Engineer and Metallurgist." *Maryland Historical Magazine* 60, 1: pp. 15–25.

ALEXANDER, JOHN HENRY. 1840. *Report on the Manufacture of Iron* (Annapolis, William McNeir).

ALLEN, ETHAN. 1860. *Clergy in Maryland of the Protestant Episcopal Church since the Independence of 1783* (Baltimore, James S. Waters).

ALLEN, JAY DOUGLAS. 1970. "The Mount Savage Iron Works, Mount Savage, Maryland. A Case Study in Pre-Civil War Industrial Development." Master's thesis, University of Maryland.

AMERICAN SOCIETY OF CIVIL ENGINEERS. 1972. *A Biographical Dictionary of American Civil Engineers* (New York, American Society of Civil Engineers).

APPLETON'S *Cyclopaedia of American Biography*, ed. by James Grant Wilson and John Fiske (New York, D. Appleton and Company, 1888).

ARMSTRONG, JAMES EDWARD. 1907. *History of the Old Baltimore Conference (Methodist)* (Baltimore, King Bros.).

ASBURY, FRANCIS. 1958. *Journal and Letters*, Elmer T. Clark, ed. (3 v., London and Nashville, Epworth Press and Abingdon Press).

BEACH, MOSES YALE. 1845. *Wealth and Biography of the Wealthy Citizens of New York City* (6th ed., New York, Sun Office).

BENNETT, GEORGE. 1927. "History of Bandon." *Oregon Hist. Quart.* **28**: pp. 311–357.

BERTHOFF, ROWLAND T. 1968. *British Immigrants in Industrial America 1790–1950* (New York, Russell & Russell, reissue).

BIDDLE, NICHOLAS. 1919. *Correspondence of Nicholas Biddle,* Reginald C. McGrane, ed. (Boston and New York, Houghton Mifflin Company).

Biographie Universelle (Michaud) (Paris, n.d.).

BIRCH, ALAN. 1968. *Economic History of British Iron and Steel Industry* (New York, Augustus M. Kelly).

BLAKE, WILLIAM J. 1849. *History of Putnam County, New York* (New York, Baker & Scribner).

BOWEN, ELI. 1854. *Off-hand Sketches* (Philadelphia, J. W. Moore).

BUCKINGHAM, J. S. [1842]. *The Eastern and Western States of America* (3 v., London, Fisher, Son & Co.).

BUSBY, C. A. 1808. *A Series of Designs for Villas and Country Houses* (London, J. Taylor).

CASWELL, A. 1837. "On Zinc as a Covering for Buildings." *Amer. Jour. Science,* First series, **31**: pp. 248–252.

CHESAPEAKE AND OHIO CANAL COMPANY. 1837. *Ninth Annual Report, June 12, 1837* (Washington, Gales and Seaton).

——. 1839. *Eleventh Annual Report, June 3, 1839* (Washington, Gales and Seaton).

CLARK, VICTOR S. 1929. *History of Manufactures in the United States* (2 v., New York, McGraw-Hill Book Company).

CLOUGH, SHEPARD B. 1946. *A Century of American Life Insurance: A History of the Mutual Life Insurance Company of New York 1843–1943* (New York, Columbia University Press).

Coal and Iron Mines of the Union Potomac Company and of the Union Company. (Baltimore, John Murphy, 1840).

CUNZ, DIETER. 1948. *The Maryland Germans* (Princeton, Princeton University Press).

DETMOLD, C. E. [1849]. *George's Creek Coal and Iron Co.* (N.p., n.n.).

Dictionary of American Biography, ed. by Allen Johnson and Dumas Malone (New York, Charles Scribner's Sons, 1928–1944).

Dictionary of National Biography (London, Oxford University Press, 1959–1960).

Dictionnaire de Biographie Française (Paris, 1967).

Dictionnaire de l'Industrie, Manufacturière, Commerciale et Agricole 5, (Paris, J. B. Bailliere, 1836–1841).

DUCATEL, J. T. 1837. "Outlines of the Physical Geography of Maryland, embracing its prominent Geological features." *Trans. Md. Acad. Science & Literature* 1 (Baltimore, published by the Academy).

——. 1838. *Annual Report of the Geologist of Maryland, 1838* (Annapolis, n.n.).

——. 1840. *Annual Report of the Geologist of Maryland, 1840* (Annapolis, n.n.).

DUFRÉNOY, OURS-PIERRE. 1837. *Voyage Métallurgique en Angleterre* (2 v., Paris, Bachelier).

ELLIS, FRANKLIN, ed. 1882. *History of Fayette County, Pennsylvania* (Philadelphia, L. H. Everts & Co.).

ELSAS, MADELEINE, ed. 1960. *Iron in the Making. Dowlais Iron Company Letters 1782–1860* (London, Glamorgan County Council).

GALE, L. D. March, 1836. "On the Uses of Zinc for Roofing of Buildings." *Mechanics' Magazine,* pp. 163–166.

——. 1837. "On Zinc Roofing." *Amer. Jour. Science,* first series, **32**: pp. 315–319.

George's Creek Coal and Iron Company. 1836. *George's Creek Coal and Iron Company* (n.p., n.n.).

——. 1839. *Untitled report of directors, 1839* (n.p., n.n.).

GOVAN, THOMAS P. 1959. *Nicholas Biddle, Nationalist and Public Banker 1786–1844* (Chicago, University of Chicago Press).

HAMILTON, THOMAS. 1968. *Men and Manners in America* [1833] *with additions from the edition of 1843* (Two volumes in one, New York, Augustus M. Kelley).

HAMMOND, BRAY. 1957. *Banks and Politics in America from the Revolution to the Civil War* (Princeton, Princeton University Press).

HARVEY, KATHERINE A. 1969. *The Best-Dressed Miners. Life and Labor in the Maryland Coal Region, 1835–1910* (Ithaca, Cornell University Press).

HILGARD, J. E. 1877. "Memoir of John H. Alexander." National Academy of Sciences, *Biographical Memoirs* 1 (Washington, National Academy of Sciences).

HONE, PHILIP. 1927. *The Diary of Philip Hone, 1828–1851,* Alan Nevins, ed. (New York, Dodd, Mead and Company).

HUGHES, GEORGE WURTZ. 1836. *Report of an Examination of the Coal Measures Belonging to the Maryland Mining Company* (Washington, National Intelligencer Office).

Inventory of the Church Archives of Maryland, Protestant Episcopal: Diocese of Maryland (Baltimore, The Maryland Historical Records Survey, 1940).

JENKS, LELAND HAMILTON. 1938. *The Migration of British Capital to 1875* (New York, Alfred A. Knopf).

JOHNSON, J. E., JR. 1917. *Blast-Furnace Construction in America* (New York, McGraw-Hill Book Company).

JOHNSON, WALTER R. 1841. *Notes on the Use of Anthracite in the Manufacture of Iron* (Boston, Charles C. Little and James Brown).

Journal of a Convention of the Protestant Episcopal Church of Maryland, 1835 (Baltimore, 1835).

Journal of a Convention of the Protestant Episcopal Church of Maryland, 1836 (Baltimore, 1836).

Journal of a Convention of the Protestant Episcopal Church of Maryland, 1837 (Baltimore, Jas. Lucas & E. K. Deaver, 1837).

Journal of a Special Convention of the Protestant Episcopal Church of Maryland 1839 (Baltimore, Lucas & Deaver, 1839).

Journal of a Convention of the Protestant Episcopal Church of Maryland, 1841 (Baltimore, Joseph Robinson, 1841).

LOUDON, JOHN C. 1835. *Encyclopedia of Cottage, Farm, and Villa Architecture* (London, Longman, Rees, Orme, Brown, Green, & Longman).

MARYLAND AND NEW YORK IRON AND COAL COMPANY. [1842]. *A Statement in Respect of the Maryland & New York Iron & Coal Company under the Management of a Committee of English Stock & Bondholders, and with English Agents and Superintendents in America* ([London], n.n.).

MARYLAND GENERAL ASSEMBLY. *Laws made and passed at sessions of the General Assembly:* 1827, 1828, 1833, 1834, 1835, 1836, 1837, 1853.

MARYLAND GENERAL ASSEMBLY, SENATE. 1838. *Journal of Proceedings, December Session, 1838.* (Annapolis).

Matchett's Baltimore Director for 1835–36 (Baltimore, R. J. Matchett, 1835).

Matchett's Baltimore Director for 1837–38 (Baltimore, R. J. Matchett, 1837).

Mitchell's Traveller's Guide through the United States (Philadelphia, Thomas, Cowperthwait & Co., 1836).

National Cyclopaedia of American Biography (New York, James T. White & Co., 1893–1919).

NEEDHAM, M. 1831. *The Manufacture of Iron* (London: n.n.).

NEW-YORK HISTORICAL SOCIETY. 1885. *Collections of the New York Historical Society for the Year 1884,* 17 (Kemble Papers) (New York, printed for the Society).

OVERMAN, FREDERICK. 1850. *The Manufacture of Iron in all its Various Branches* (Philadelphia, H. C. Baird).

——. 1882. *A Treatise on Metallurgy* (6th ed., New York, D. Appleton and Company).

PALMER, HENRY ROBINSON. 1824. *Description of a Railway on a New Principle* (2d ed., revised, London, printed for J. Taylor).

PEARRE, NANCY C. 1964. "Mining for Copper and Related Minerals in Maryland." *Md. Hist. Mag.* **59**, 1: pp. 15–33.

PESSEN, EDWARD. 1970. "The Wealthiest New Yorkers of the Jacksonian Era: A New List." *New-York Hist. Soc. Quart.* **54**: pp. 145–172.

PINKNEY, WILLIAM. 1867. *Memoir of John H. Alexander, Ll.D.* (Baltimore printed for Md. Hist. Soc. by J. Murphy).

POCOCK, WILLIAM F. 1807. *Architectural Designs for Rustic Cottages, Picturesque Villas, &c.* (London, printed for J. Taylor).

PREUSS, CHARLES. 1958. *Exploring with Frémont*, Trans. and ed. by Erwin G. and Elisabeth K. Gudde (Norman, University of Oklahoma Press).

PROUD, JOHN G. 1868. *Memoirs of Deceased Alumni of St. John's College, Annapolis* (Baltimore, Wm. K. Boyle).

RANKIN, ROBERT G. 1855. *The Economic Value of the Semi-Bituminous Coal of the Cumberland Coal Basin* (New York, John F. Trow).

REDLICH, FRITZ. 1968. *The Molding of American Banking. Men and Ideas* (New York and London, Johnson Reprint Corporation).

RICE, OTIS K. 1970. *The Allegheny Frontier. West Virginia Beginnings, 1730–1830* (Lexington, University Press of Kentucky).

RIDEING, WILLIAM H. 1879. "The Old National Pike." *Harper's New Monthly Magazine* **59**, pp. 801–816.

SANDERLIN, WALTER S. 1946. *The Great National Project: A History of the Chesapeake and Ohio Canal*, Johns Hopkins University Studies in Historical and Political Science, Series LXIV, No. 1 (Baltimore, Johns Hopkins Press).

SCHARF, J. THOMAS. 1882. *History of Western Maryland* **2**, (Philadelphia, L. H. Everts).

SCHLESINGER, ARTHUR M., JR. 1950. *The Age of Jackson* (Boston, Little, Brown and Company).

SCHUBERT, H. R. 1957. *History of the British Iron and Steel Industry from c. 450 B.C. to A.D. 1775* (London, Routledge & Kegan Paul).

SCOVILLE, JOSEPH A. 1862–1866. *The Old Merchants of New York City* (5 v., New York, Carleton).

SCRIVENOR, HARRY. 1854. *History of the Iron Trade* (London, Longman, Brown, Green and Longman).

SEARIGHT, THOMAS B. 1894. *The Old Pike, A History of the National Road* (Uniontown, Pa., published by the author).

SILLIMAN, BENJAMIN. 1838. *Extracts from a Report Made to the New York and Maryland Coal & Iron Company* (London, T. C. Savill).

SINGEWALD, JOSEPH T., JR. 1911. *Report on the Iron Ores of Maryland with an Account of the Iron Industry*, Md. Geological and Economic Survey (Special Publication 9, 3). (Baltimore, Johns Hopkins Press).

SMITH, WALTER B., and ARTHUR H. COLE. 1935. *Fluctuations in American Business 1790–1860* (Cambridge, Mass., Harvard University Press).

SWANK, JAMES M. 1878. *Introduction to a History of Iron-making and Coal Mining in Pennsylvania* (Philadelphia, published by the author).

——. 1892. *History of the Manufacture of Iron in All Ages* (2d. ed., Philadelphia, American Iron and Steel Association).

TAUSSIG, FRANK W. 1931. *The Tariff History of the United States* (8th ed., New York, G. P. Putnam's Sons).

THOMAS, JAMES WALTER, and T. J. C. WILLIAMS. 1923. *History of Allegany County* ([Cumberland, Md.], L. R. Titsworth & Company).

TOMLINSON, CHARLES. ed. 1854. *Cyclopedia of Useful Arts & Manufactures.* (2 v., London and New York, G. Virtue & Co.).

TREDGOLD, THOMAS. 1825. *A Practical Treatise on Rail-Roads and Carriages* (London, printed for Josiah Taylor).

TROLLOPE, FRANCES. 1960. *Domestic Manners of the Americans*, Donald Smalley. ed. (New York, Vintage Books).

TYSON, PHILIP T. 1837. "A Description of the Frostburg Coal Formation of Allegany County, Maryland, with an account of its geological position." *Trans. Md. Acad. Science and Literature* **1** (Baltimore, published by the Academy).

——. 1837. "A Descriptive Catalogue of the Principal Minerals of the State of Maryland." *Trans. Md. Acad. Science and Literature* **1** (Baltimore, published by the Academy).

VAN WAGENEN, JARED, JR. 1954. *The Golden Age of Homespun* (Ithaca, Cornell University Press).

VARLE, CHARLES. 1833. *A Complete View of Baltimore* (Baltimore, Samuel Young).

VIRGINIA BOARD OF PUBLIC WORKS. 1832. *16th Annual Report, Feb. 7, 1832* (Richmond, Samuel Shepherd & Co.).

——. 1833. *17th Annual Report, Jan. 17, 1833* (Richmond, Samuel Shepherd & Co.).

——. 1834. *18th Annual Report, Dec. 18, 1833* (Richmond, Samuel Shepherd & Co.).

——. 1835. *19th Annual Report, Feb. 3, 1835* (Richmond, Samuel Shepherd & Co.).

——. 1837. *21st Annual Report, Jan. 26, 1837* (Richmond, Samuel Shepherd & Co.).